ANTICIPATIONS

of the

Reaction of Mechanical and Scientific Progress Upon Human Life and Thought

H. G. WELLS

DOVER PUBLICATIONS, INC.
Mineola, New York

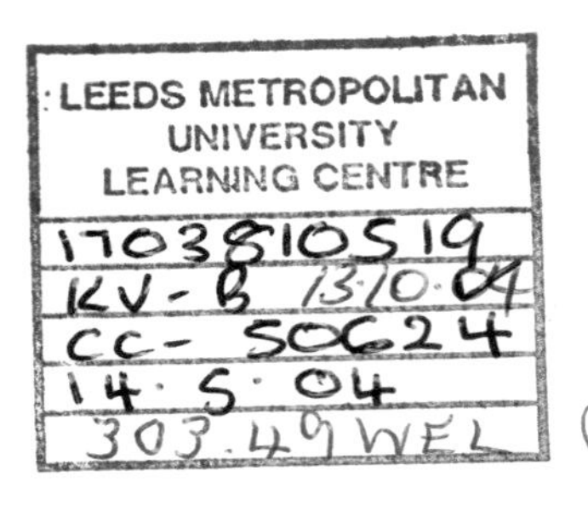

Copyright

Bibliographical Note

This Dover edition, first published in 1999, contains the complete text of *Anticipations of the Reaction of Mechanical and Scientific Progress Upon Human Life and Thought,* as first published in 1902 by Bernhard Tauchnitz, Leipzig. Included are the author's preface to a later (1914) edition and a new introduction by Martin Gardner, specially prepared for this edition.

Library of Congress Cataloging-in-Publication Data

Wells, H. G. (Herbert George), 1866–1946.
Anticipations of the reaction of mechanical and scientific progress upon human life and thought / H. G. Wells.
p. cm.
ISBN 0-486-40673-3 (pbk.)
1. Social prediction. I. Title.
HN16.W44 1999
303.49–dc21 99-14038
CIP

Manufactured in the United States of America
Dover Publications, Inc., 31 East 2nd Street, Mineola, N.Y. 11501

INTRODUCTION TO THE DOVER EDITION

by Martin Gardner

HERBERT George Wells (1866–1946) was 35 in 1901 when this book was first published in Leipzig by Bernhard Tauchnitz. Titled *Anticipations of the Reaction of Mechanical and Scientific Progress Upon Human Life and Thought,* it had earlier been serialized in *The Fortnightly Review* (April through December 1901) with the subtitle: *An Experiment in Prophecy.* Wells was already famous for his science fiction novels and short stories. A second edition of *Anticipations* in 1902 contained numerous revisions and additions. In 1914 Wells added a new introduction, here reprinted, to a cheaper edition of the book.

Anticipations was Wells's first best-seller. The book had an enormous impact on British intellectuals and their European counterparts. George Bernard Shaw, Sidney and Beatrice Webb, William and Henry James, and Arnold Bennett were among a raft of eminent writers who highly praised the book.

Through his life Wells fancied himself a shrewd prophet of events to come. Many of his later books and articles (such as *What Is Coming?* (1914), *A Year of Prophesying* (1924); *The Way the World Is Going* (1928); *The Shape of Things to Come* (1933); and *The Outlook for Homo Sapiens* (1942) were similar to *Anticipations* in their efforts to predict the future.

Wells also explored the future in his science fiction. His wildest misses were novels and tales about Martians, and a novel about intelligent humanoids living in caves below the surface of the Moon. Several utopian fantasies, and one negative utopian novel (a nightmare vision of a possible future) are mixtures of hits and misses. It was in his 1914 novel *The World Set Free* that

Wells made his most astounding hit. Dedicated to Frederick Soddy for his pioneer research on radium, the novel opens with a moving extract from the diary of a physicist who has found a way to split the atom and release atomic energy. He is fearful of the consequences of his discovery, but realizes that, had he not made it, other physicists soon would. The novel describes a war between England and Germany, in the middle of the twentieth century, during which "atomic bombs," as Wells actually called them, were dropped from airplanes.

Wells was always good at seeing the immediate consequences of new technologies, but science has a wonderful way of springing great surprises. On this account, we should not fault Wells for never anticipating, in this or any other book, the rapid development of motion pictures and television, the computer revolution, the DNA revolution, how soon astronauts would land on the Moon, or how soon space probes would be exploring our solar system.

In the first chapter of *Anticipations* Wells scored a number of hits. He predicted that coal and steam, as power sources for locomotives and ships, would soon be replaced by what he called "explosion engines" driven by gas and oil. He suggested that the piston might some day be replaced by a rotary device such as the one now used in Mazdas. He expected electrical cars to become practical. He foresaw the coming of huge trucks in competition with railroads, and the replacement of horse-drawn vehicles by cars and omnibuses.

The rapid proliferation of trucks and cars, Wells realized, would require new and wider roads, the dirt replaced by asphalt or some other hard substance. The roads would be shaped to drain rainfall, and have guardrails to prevent cars from plunging off embankments. He predicted bridges and underpasses where highways intersected, and the need for strong traffic regulations.

Train rides, Wells foresaw, would soon be smooth enough to allow dining. His suggestion that the century would see moving sidewalks, running side by side at different speeds, has failed to materialize except in the case of some large airline terminals. Wells's greatest miss in this chapter is his failure to anticipate how rapidly air travel would arrive. "I do not think it at all probable," he says in a footnote, "that aeronautics will ever come into play as a serious modification of transport and communication. . . . Man is not, for example, an albatross, but a land biped. . . . "

Wells next turns his attention to the Earth's monstrous cities. Borrowing terms from physics, he considers the centripetal forces causing congestion, and the centrifugal forces sending city dwellers to the suburbs. He rightly guesses that trains and cars will intensify this dispersion, with telephones also playing a role. Unfortunately, Wells got carried away with his estimation of the dominance of centrifugal forces. The great cities, he predicted, "are destined to such a process of dissection and diffusion as to amount almost to obliteration. . . ." By the year 2000, he writes, it is probable that London's "urban region" would include almost all of England and parts of Wales. In Eastern United States, he predicted that the "vast stretch of country" from Washington, D.C., to Albany, would become home for citizens working in New York City and Philadelphia.

Wells hit the mark in sensing that increasing numbers of business firms would move from cities to uncrowded towns and suburbs. Alas, what he calls the "pauper masses" still huddle in most of the world's metropolises; the thousands of sick and homeless creating social problems as seemingly unsolvable today as they have been in the past.

Chapter 3, on changing social conditions, predicts the rise in advanced nations of several classes new to history: There will be stockholders who "do nothing in common except receive and hope for dividends". There will also be a class made up of the working poor, whom Wells calls "a multitude of people drifting down towards the abyss," victims of new technologies for which they are untrained. Wells sees this group being replaced by machines or thrown out of work by the flight of companies to lands where labor is cheaper. Finally, Wells expects a vast, chaotic, educated middle class of professional specialists.

Wells realized that the rapid progress of science would alter the social fabric of advanced cultures in unpredictable ways. Chapter 3's chief value is its vivid portrayal of British life at the time of its writing. Two gems are an hilarious footnote describing a meeting of the House of Commons, and Wells's account of the bumbling process by which one of his houses was built.

It is hard to fault most of the predictions in Chapter 4. Wells describes a typical middle class home of the future as one surrounded by a yard, and centrally heated by warm air blown from wall ducts. Oil lamps, he expects, will be replaced by electric lights. Stoves will be electric. Chimneys will either vanish or remain rising from bogus fireplaces with fake glowing logs.*

**Parade* (November 15, 1998) ran a half-page ad for "The Ultimate Fireplace Video." For six hours the TV screen shows a picture of burning, crackling logs. "It's soothing," reads the ad. "It's romantic. And it's maintenance free . . . no chopping wood, no fancy equipment to by. No cinders and soot to clean. And no smoky living room." And, of course, no heat.

(Wells stumbled in anticipating fake smoke rising from the sham chimneys.) Every bedroom will have an adjoining bathroom. Men will no longer feel obliged to wear boots indoors.

Rich shareholders–Wells calls them the "leisure class"–will control architecture, art, and fashion. Striking new styles of clothing will come and go. Popular novels and plays will reflect the doings of the leisure class. There will be nursery schools for the very young.

Moral restrictions will decline as men and women seek greater sexual freedom. Divorces will increase. There will be more childless marriages, more children born out of wedlock. Regions of "opulent enjoyment," like the French Riviera resorts of Wells's day, will flourish. The loss of moral and religious certainty will create great confusion and unrest until stable patterns of ethics emerge unencumbered by stale religious creeds.

Two whimsical misses are worth noting in this chapter. Wells suggests that windows of the future will be cleaned by jets of soapy water flowing from holes in a horizontal pipe above each window. To ease the cleaning of floors he expects sharp angles between floors and walls to be covered by curved surfaces.

Wells's fifth chapter contains the first of many attacks he was to make on what he sees as degenerate forms of democracy responsible for poor government and terrible wars. Wells had no respect for a system in which power is in the hands of ignorant voters easily swayed by demagogs. There is, he insists, "nothing in the mind of the average man except blank indifference." For the rest of his life Wells hoped and believed that mass democracy would be replaced by a socialist government in the hands of an aristocracy of well educated, scientifically minded men. They would take power, Wells believed, like butterflies emerging from ugly cocoons.

Wells never lost his enthusiasm for a world state ruled by such an elite. In later books he gave this ruling class such names as the "Samurai" (in *A Modern Utopia,* 1904), "Open Conspirators" (in *The Open Conspiracy,* 1928), and the "Air Police" (in *The Shape of Things to Come,* 1933). Wells's Samurai echo the Guardians in Plato's *Republic,* the first great Utopian and a work Wells greatly admired. Samurai clubs sprang up here and there in England. Wells even made an abortive effort to transform the socialist Fabian Society, then controlled by Sidney and Beatrice Webb, into an organization of "Open Conspirators."

For a short time Wells viewed Lenin's revolution in Russia as not far from his notion of a great state taken over by an efficient elite. When he visited Russia and met Lenin–a visit he recorded in *Russia in The Shadows* (1920)–he found fault with many aspects of communism, but there is no hint that he deplored its total absence of democracy. Indeed, to put it bluntly, the world state outlined here and in *A Modern Utopia* is a police state. Wells never made it clear whether his Samurai would take power gradually or by a bloody revolution. It never occurred to him, one regrets to say, that intelligent, science-trained leaders, might be just as willing to work for an evil dictator as for a democratic society that respected free speech and human rights. It was the same mistake that would be made years later by those who called themselves technocrats. Wells seemed never to ask himself why his Samurai, once they controlled a nation, should be more concerned for the welfare of all than the welfare of themselves alone.

In later books and articles Wells would elaborate at length on what he believed was the inevitable emergence of a world state governed by enlightened technocrats. What of the possibility that in the aftermath of wars great nations would fall into the hands of ruthless Napoleons? "Nothing of the sort is going to happen," Wells declares in this book. "The day of individual leaders is past."

As he grew older and wiser, Wells moderated his early attacks on representative democracies in which all citizens have a right to vote. In *The Work, Wealth, and Happiness of Mankind* (1931) he suggests ways of improving democracy by such means as proportional representation, and the elimination of two legislative bodies, in England and America for instance, when a single body would be sufficient. He sees the possibility of civil servants running governments, but admits that "no one has yet worked out any better way [than democracy] of getting general assent to administration and legislation. A much better way cannot be beyond human contriving, but it has not yet been contrived."

Wells also was never clear on exactly how nations would come together to create a United States of the World governed by a science-trained aristocracy that would combine some sort of representative democracy with economic socialism. Was his vision of a worldwide utopia, free forever of wars and gross injustices, a genuine possibility? Or was it no more than a dream?

All his life Wells alternated between moods of hope and moods in which he feared such a world state would never come to pass. As he approached death, a profound pessimism overcame him. He saw no signs of movement towards a world government.

There were only intensifying nationalisms, and hideous wars waged with ever more powerful weapons of mass destruction. Such hopelessness tinged *The Fate of Homo Sapiens* (1939), and culminated in the black despair of *The Mind at The End of Its Tether* (1945). This sad little book was written shortly before Wells died in 1946, but not before he read that atomic bombs, which he had imagined and named thirty years earlier, had fallen on Japan.

Wells's vision of future warfare, the substance of Chapter 6 in *Anticipations,* bristles with flawed predictions. Although he sees a role in modern warfare for motor cars and tanks, he calls tanks "iron tortoises," too slow moving to be effective. Modern soldiers, he believes, will move rapidly on bicycles! Rifles, of course, will be improved; guns will have longer ranges. Huge balloons, armed with men, guns, and perhaps bombs, will fight awesome air battles. (Wells devotes several lurid pages to describing such scenes.) And tethered balloons, carrying telescopes, he says, will serve as "argus eyes" to spy on fields ahead, illuminating night landscapes with huge searchlights.

Sea battles will be fought by small, swift ships armed with guns and battering rams. Such battles will be brief. "The struggle between two naval powers on the high seas . . . will not last more than a week or so." Submarines? "I must confess," Wells writes, "that my imagination, in spite even of spurring, refuses to see any sort of submarine doing anything but suffocate its crew and founder at sea." Their torpedoes, he assures his readers, will have as much chance of hitting a target as a man "blindfolded, turned around three times, and told to fire revolver-shots at a charging elephant."

The chapter scores a few hits. The winning side in the modern war, Wells foresees, will be the side with the best educated fighting men, and the best scientists and engineers behind the lines. He correctly perceives that during a great war, an entire nation must turn socialist, its government taking control of every aspect of the economy. He anticipates many useless wars in which young men will be slaughtered for no rational cause.

"Tramp, tramp, tramp, they go, boys who will never be men, rejoicing patriotically in the nation that has thus sent them forth, badly armed, badly clothed, badly led, to be killed in some avoidable quarrel by men unseen." This early in the century, not even Wells could conceive of the deaths of tens of thousands of civilians from bombs dropped from planes. As we know, such senseless wars still rage around the world. Such wars will never cease, Wells was convinced, until there is a viable world state.

Nor would a world government emerge, Wells believed, until a worldwide language did. The three major contenders, he writes in Chapter 7, are English, French, and German. As other languages, with the exception of Asian ones, fade, Wells sees the western nations slowly becoming bilingual. French will dominate in the short run, but in the long run English will prevail. German is dismissed as too "unattractive, unmelodious, and cursed with a hideous lettering. . . ."

Most of the economic and political arguments of Chapter 8 are now hopelessly obsolete. The chapter's central theme is the sluggish movement toward a single government that Wells calls the New Republic. The movement will accelerate as nations become more intertwined economically and culturally, as English becomes the world's dominant language, and as more books are translated into other languages. Wells bemoans the absence of intelligent debates on major issues. He imagines how wonderful it would be if important controversial books were annotated by writers who held contrary opinions. He longs for a general index of all books on sale, listed alphabetically by title and by author. Such an index, though only for American books, is now available in hard cover, and on computer screens as the many-volumed *Books in Print.*

Wells's final chapter swarms with views he would later regret. In it, he reveals that although he considers the omniscient, all-powerful deity of Christianity absurd, he does not want to discard God altogether. In place of the Judeo-Christian God, with infinite attributes, Wells proposes a finite God, unrelated to any established faith. The elite who rule the New Republic will worship such a God, he suggests, recognizing that He is indeed transcendent, as far beyond our understanding as our world is beyond the grasp of an amoeba. Somehow, such a deity will provide a meaning for the universe and human history, even though we cannot know what the meaning is.

Wells seems unaware that this concept of a finite God had earlier been put forth by many philosophers. He would argue again for such a deity in his World War I novel *Mr. Britling Sees It Through,* and in two 1917 works, *God: the Invisible King,* and *The Soul of a Bishop.* Later he would repudiate the concept altogether, and declare himself an honest atheist. Today, a finite God is defended by the so-called process theologians, notably Charles Hartshorne, who believe God exists in time, and is evolving as His creation evolves.

Throughout his life Wells vigorously defended a belief in free will. He does not mention Kant, but his approach to the free will problem is identical with Kant's. In the world of science, Wells writes in his last chapter, there is strict cause-and-effect determinism. But the human will is somehow outside the "rigidly predestinate" world of "atoms and vibrations." It is free "just as new-sprung grass is green, wood hard, ice cold, and toothache painful." Wells believes that by exercising free will, humanity will finally construct a utopia of peace and justice. The old religious ethics will give way to a new set of morals, though the posits on which such a morality will rest are never made clear.

In the decades before the rise of a world state, Wells predicts that Protestant Christianity will slowly decay, to be replaced, as the world's dominant faith, by Roman Catholicism. Those who refuse to follow this trend, Wells warns, will turn to weird pseudoscientific cults such as theosophy, Spiritualism, and Eastern religions. Rich young men and women will dabble in witchcraft and devil worship just for the fun of it. On this score, of course, Wells's crystal ball was quite accurate. Gilbert Chesterton is thought to have said somewhere that when people stop believing in God, they believe anything. Wells here says it this way: "The fool hath said in his heart, 'there is no God,' and after that he is ready to do anything with his mind and soul."

Wells's passing interest in a finite God was a harmless diversion, but there is a dark side to this chapter: It defends an extreme program of negative eugenics. The New Republic leaders, Wells writes, "favour the procreation of what is fine and efficient and beautiful in humanity–beautiful and strong bodies, clear and powerful minds, and a growing body of knowledge–and to check the procreation of base and servile types . . . of all that is mean and ugly and bestial in the souls, bodies, and habits of men."

Just how is this to be done? By mercy killings! The leaders of the new world will have "little pity and less benevolence." They "will not be squeamish" about inflicting death on the unfit "because they will have a fuller sense of the possibilities of life than we possess. They will have an ideal that will make killing worth the while; like Abraham, they will have the faith to kill. . . ."

And who are the unfit that are to be thrown away? Wells ticks them off: those with transmittable diseases, with mental disorders, with bodily deformations, the criminally insane, even the incurable alcoholic! All are to be put to death humanely–by first giving them opiates to spare them needless suffering!

How will the New Republic handle what Wells calls "inferior races"? He asks: "How will it deal with the black? How will it deal with the yellow man? How will it tackle that alleged termite in the civilized woodwork, the Jew?"

After the world state is in place, perhaps not until after 2000, Wells foresees that sterilization, killing, and birth control methods will effect a gradual fading from the Earth of inferior races. "There is something very ugly about many Jewish faces," Wells writes, then he quickly adds, "there are Gentile faces just as coarse and gross." Many Jews "are intensely vulgar in dress and bearing, materialistic in thought, and cunning and base in method, but no more so than many Gentiles."

Yet Wells sees no reason for any special effort to eliminate Jews. Those who have a tendency toward "social parasitism" will be treated just like similar Caucasians. Increasing intermarriages of Jews and Gentiles, Wells predicts, will be sufficient to cause Jews to "cease to be a physically distinct element in human affairs in a century or so." As for "those swarms of blacks, and brown, and dirty-white, and yellow people," who do not meet the needs of the New Republic, "they will have to go. . . . So far as they fail to develop sane, vigorous, and distinctive personalities for the great world of the future, it is their portion to die out and disappear."

From our perspective, of course, Wells's statements about inferior races, and the use of killing as a tool to weed out the unfit, come perilously close to Hitler's efforts to breed a superior Aryan race, and to "solve the Jewish question" with the aid of gas chambers. Nor do we know whether Wells ever apologized for this portion of his last chapter. In his autobiography (1934) he calls *Anticipations* the "keystone to the main arch of my work." However, as early as 1905, in *A Modern Utopia* (Chapter 5) Wells presents a much less harsh defense of eugenics. He still advises ridding the world of the unfit, but now the only means he proposes is sterilization. "So soon as there can be no doubt of the disease or baseness of the individual, so soon as the insanity or other disease is assured, or the crime repeated a third time, or the drunkenness or misdemeanor past its seventh occasion (let us say), so soon must he or she pass out of the common ways of men. . . ." There will be incarceration and sterilization, but "no killing, no lethal chambers."

There is a section on eugenics in a much later work, *The Work, Wealth, and Happiness of Mankind* (1931). Here Wells still purports to favor isolating and sterilizing the unfit, but he now clearly recognizes that human nature is far too complex to justify any controlled breeding of superior bodies and minds. "We do not want human beings to become simply taller or swifter or web-footed or what not. We want a great variety of human beings. . . . We must tolerate much that is odd and weak lest we lose much that is glorious and divine. . . . For many generations, and perhaps for long ages, we must reckon with a population of human beings not very different from those we have to deal with today. . . . It is to a better education and to a better education alone, therefore, that we must look for any hope of ameliorating substantially the confusions and distresses of our present life."

In *Anticipations'* final chapter, Wells suggests that although leaders of the New Republic will believe they are serving an unknowable God, this quasi-religious faith will not include a belief in immortality. There may be an afterlife, Wells admits, but "on this side, in this life," there is no evidence that our "egotisms" will survive death. Citizens of the world state will replace the notion of heaven with the idea of a bright future for humanity. "For that future these men will live and die." Why they will wish to live and die, especially to sacrifice their life, for a future they will never see, is a question Wells never adequately answers.

Surprisingly, Wells made no changes in the text of *Anticipations'* 1914 edition. He did add a new introduction in which he confesses that the book contains "several rash and harsh generalizations," but on the whole he is pleased with "how little there is in it that I would change were I to rewrite it. . . ." His chief apology is for a failure to foresee how rapidly aviation would develop, and for his "very stale . . . anticipations of aerial war." He makes no apology for his "onslaught" on mass democracy, or for his vision of a world state governed by an enlightened, science-minded elite. "It is my faith," he writes. "It is my form of political thought."

AN INTRODUCTION TO THE 1914 EDITION

IT is now nearly fifteen years since *Anticipations* was written, and it is with a certain detachment and curiosity that I have read it over again to consider very seriously whether the issue of a fresh edition is justifiable. I have looked at the book only very occasionally since its first publication, I have never read it through since I passed the proofs for press until the present occasion, and on the whole I am surprised to find how little there is in it that I would change were I to rewrite it at the present time. It is a better book than I have been in the habit of thinking it was, and whatever the value of both of them to the world at large may be, the H. G. Wells of thirty-three has little to be ashamed of in presenting his book to the criticisms of the H. G. Wells of forty-eight. There are places, as I will presently indicate, which the latter, with some advantages of travel and experience, may be inclined to consider a little thin; there are ignorances and there are several rash and harsh generalizations; but an occasional trick of harshness and moments of leaping ignorance are in the blood of H. G. Wells; everybody who reads him has to stand that–he has to stand it himself more than any one–and forty-eight has, I fear, but little reason on that score for a superior attitude to thirty-three. It is like a lisp or an ugly voice. On the whole, and that is the astonishing thing, the book stands; there are places when you might very well think the writer was writing about the present instead of lunging boldly into what was then the future; and it may even be that in checking its forecasts by accomplishment, the reader will find an interest that his predecessor at the beginning of the century necessarily lacked.

Remember that the book was written during the Boer War, and before the complete publication of the 1901 census returns.

Since then not only these but the returns of 1911 have come to hand to confirm very thoroughly the anticipations of urban extension, of social segregation and of the altering weight of classes that constitutes the opening chapters. All that has worked out very satisfactorily, so that even quite detailed prophecies have been confirmed, such as the disappearance of literary "Boomsters" [pp. 78–79], and the rot of the press and the appearance of smaller newspapers [. . .]. And further on a considerable claim for verification may be based upon the estimate of Russia's power, made five years before the revolt of Moscow and the war with Japan. The whole of that chapter, the Larger Synthesis, has stood the wear of fourteen years remarkably well. For the most part it might have been written yesterday. But on the other hand, there are undeniable failures. Those specialized roads for motors, for example, and particularly the one that was to run from London to Brighton, do not materialize, and the book displays a remarkable want of confidence in the immediate practicability of either flying machines or submarines. Almost everyone who reads this book now will laugh at my timid little bladder-assisted aeroplanes, and yet, in 1901, I was considered a very extravagant young man. "Long before 2000, and very probably before 1950, a successful aeroplane"–the boldness of it! The very stalest part[s] of *Anticipations* are the anticipations of aerial war. But the laugh in that matter is more against me than the uninformed would believe, for even as I wrote these hesitating words, there lay in the bureau at which I wrote a pile of notes upon aviation, which a certain young soldier had confided to my keeping before he went to South Africa. He had come to me because I, at anyrate, did not "think the whole blessed thing idiotic." If he came back I was to return them to him, it was his secret and he would go on with it; if he was killed I was to get them published. And now the Dunne self-balancing aeroplane defies the gales, and the other day, by Captain Dunne's kindness, I was soaring three thousand feet over the town of Sheerness.

The stuff about the "New Republic," and the attempt to define the social classes of the new age, is, I think, the most permanently valuable part of this book. The general idea of the "New Republic," the onslaught on "Democracy," the manifest dislike for such partizan and particularist things as trade unionism and nationalism are as much a part of me as those intonations of my voice or the shape of my nose. That conception of an open conspiracy of intellectuals and wilful people against existing institutions and existing limitations and boundaries is always with me; it is my King Charles's head, and it forms the substance of the longest novel I have ever written–that is, if ever the war will let me get it written–the novel I am still writing. I admit that after fourteen years this open conspiracy still does not very definitely realize itself, but in that matter I have a constitutional undying patience. That open conspiracy will come. It is my faith. It is my form of political thought.

Since *Anticipations* was written I have been through the Fabian Society, and it is amusing in this moment of retrospect to recall that plunge and that tumultuous emergence. In the days when I wrote *Anticipations* I knew scarcely more of the Fabian Society than I did of the Zetetic Society, but the publication of that book and its follower *Mankind in the Making,* brought Mr. and Mrs. Sidney Webb into my world. They appeared riding very rapidly upon bicycles, from the direction of London, offering certain criticisms of my general forecast and urging me to join and stimulate the Fabians. This extraordinary couple, so able and energetic, so devoted, so perplexingly limited, exercised me enormously. Their essential criticism of *Anticipations* was that I did not sufficiently recognize the need and probability of a specialized governing class, and they expounded to my instinctively shrinking intelligence that conception of a great bureaucracy which it has been their life-work to convey to the English intelligence. They tried to bring my New Republic within the official dimensions of their bureaucratic state while I as earnestly tried to relax their outlook to the demands of my own temperament. *A Modern Utopia* with its Samurai was the fruit of this transitory and never entirely harmonious marriage of minds, and then, recoiling as it were, I set myself with what I now perceive was an entirely exaggerated and unnecessary horror to release the Fabian Society and British Socialism from their influence. I failed scandalously after preposterous wranglings at Clifford's Inn and Essex Hall, wranglings in which Mr. Bernard Shaw somehow contrived to take a leading and entirely incomprehensible part, and which I still find too amusing to regret, and when I did at last draw breath on the further side of these discussions it was with Mr. and Mrs. Sidney Webb beyond the reach of any ordinary apology, and a much clearer, if perhaps not materially different, conception of the underlying forces of government than those set forth in this book.

I saw then what hitherto I had merely felt that there was in the affairs of mankind something unorganized which is greater than any organization. This unorganized power is the ultimate sovereign in the world. It is a thing of the intellectual life and it is also a thing of the will. It is something transcending persons just as physical or biological science or mathematics transcends persons. It is a racial purpose to which our reason in the measure of its strength, submits us. It is what was intended when people used to talk about an Age of Reason, it was vaguely apprehended when the Victorians spoke of Public Opinion. Since writing *Anticipations* I have got into the habit of using for it the not very elegant phrase, the Collective Mind. I hope someone will soon find a better expression. This Collective Mind is essentially an extension of the spirit of science to all human affairs, its method is to seek and speak and serve the truth and to subordinate oneself to one's conception of a general purpose. Its immediate social and political effect is an insistent demand for perfect freedom of thought and discussion. Social and political order it values only as a means of freedom. But in these earlier books and until I had come into contact with those dreams of official controls, "governing classes" and the like, in action, it is manifest how little I apprehended the danger of interference and paralysis which through the self-sufficiency of governing and managing persons any attempt to organize this collective mind involves. That chiefly is what I should alter if I were to rewrite *Anticipations* now. I should point out that the New Republic is not a type and a class of persons but *a power in men's minds and in mankind.* And that the worst enemies the Collective Mind can have, are a swarm of busy little bureaucrats professing to direct or protect it, gaining a kind of stifling control of it and working in its name. Order is a convenience, but Anarchism is the aim and outcome of that convenience. For the material securities of life indeed we want police and roads and maps and market rules, "efficiency" and government, but for the supreme things we have to abandon the methods of self-preservation and get out of cliques, Academies, securities and all associations. It is not by canvassing and committees, by tricks and violence, but by the sheer power of naked reasonableness, by propaganda and open intention, by feats and devotions of the intelligence, that the great state of the future, the world state, will come into being.

H. G. WELLS.

CONTENTS

ANTICIPATIONS

I. LOCOMOTION IN THE TWENTIETH CENTURY

IT is proposed in this book to present in as orderly an arrangement as the necessarily diffused nature of the subject admits, certain speculations about the trend of present forces, speculations which, taken all together, will build up an imperfect and very hypothetical, but sincerely intended forecast of the way things will probably go in this new century.* Necessarily diffidence will be one of the graces of the performance. Hitherto such forecasts have been presented almost invariably in the form of fiction, and commonly the provocation of the satirical opportunity has been too much for the writer;** the narrative form becomes

*In the earlier papers, of which this is the first, attention will be given to the probable development of the civilised community in general. Afterwards these generalisations will be modified in accordance with certain broad differences of race, custom, and religion.

**Of quite serious forecasts and inductions of things to come, the number is very small indeed; a suggestion or so of Mr. Herbert Spencer's, Mr. Kidd's *Social Evolution,* some hints from Mr. Archdall Reid, some political forecasts, German for the most part (Hartmann's *Earth in the Twentieth Century,* e.g.), some incidental forecasts by Professor Langley (*Century Magazine,* December, 1884, e.g.), and such isolated computations as Professor Crookes' wheat warning, and the various estimates of our coal-supply, make almost a complete bibliography. Of fiction, of course, there is abundance: *Stories of the Year 2000,* and *Battles of Dorking,* and the like–I learn from Mr. Peddie, the bibliographer, over one hundred pamphlets and books of that description. But from its very nature, and I am writing with the intimacy of one who has tried, fiction can never be satisfactory in this application. Fiction is necessarily concrete and definite; it permits of no open alternatives; its aim of illusion prevents a proper amplitude of demonstration, and modern prophecy should be, one submits, a more and more of a nuisance as the speculative inductions become sincerer, and here it will be abandoned altogether in favour of a texture of frank inquiries and arranged considerations. Our utmost aim is a rough sketch of the coming time, a prospectus, as it were, of the joint undertaking of mankind in facing these impending years. The reader is a prospective shareholder–he and his heirs–though whether he will find this anticipatory balance-sheet to his belief or liking is another matter.

For reasons that will develop themselves more clearly as these papers unfold, it is extremely convenient to begin with a speculation upon the probable developments and changes of the means of land locomotion during the coming decades. No one who has studied the civil history of the nineteenth century will deny how far-reaching the consequences of changes in transit may be, and no one who has studied the military performances of General Buller and General De Wet but will see that upon transport, upon locomotion, may also hang the most momentous issues of politics and war. The growth of our great cities, the rapid populating of America, the entry of China into the field of European politics are, for example, quite obviously and directly consequences of new methods of locomotion. And while so much hangs upon the development of these methods, that development is, on the other hand, a process comparatively independent, now at anyrate, of most of the other great movements affected by it. It depends upon a sequence of ideas arising, and of experiments made, and upon laws of political economy, almost as inevitable as natural laws. Such great issues, supposing them to be possible, as the return of Western Europe to the Roman communion, the overthrow of the British Empire by Germany, or the inundation of Europe by the "Yellow Peril," might conceivably affect such details, let us say, as doorhandles and ventilators or mileage of line, but would probably leave the essential features of the evolution of locomotion untouched. The evolution of locomotion has a purely historical relation to the Western European peoples. It is no longer dependent upon them, or exclusively in their hands. The Malay nowadays sets out upon his pilgrimage to Mecca in an excursion steamship of iron, and the immemorial Hindoo goes a-shopping in a train, and in Japan

branch of speculation, and should follow with all decorum the scientific method. The very form of fiction carries with it something of disavowal; indeed, very much of the Fiction of the Future pretty frankly abandons the prophetic altogether, and becomes polemical, cautionary, or idealistic, and a mere footnote and commentary to our present discontents.

and Australasia and America, there are plentiful hands and minds to take up the process now, even should the European let it fall.

The beginning of this twentieth century happens to coincide with a very interesting phase in that great development of means of land transit that has been the distinctive feature (speaking materially) of the nineteenth century. The nineteenth century, when it takes its place with the other centuries in the chronological charts of the future, will, if it needs a symbol, almost inevitably have as that symbol a steam-engine running upon a railway. This period covers the first experiments, the first great developments, and the complete elaboration of that mode of transit, and the determination of nearly all the broad features of this century's history may be traced directly or indirectly to that process. And since an interesting light is thrown upon the new phases in land locomotion that are now beginning, it will be well to begin this forecast with a retrospect, and to revise very shortly the history of the addition of steam travel to the resources of mankind.

A curious and profitable question arises at once. How is it that the steam locomotive appeared at the time it did, and not earlier in the history of the world?

Because it was not invented. But why was it not invented? Not for want of a crowning intellect, for none of the many minds concerned in the development strikes one–as the mind of Newton, Shakespeare, or Darwin strikes one–as being that of an unprecedented man. It is not that the need for the railway and steam-engine had only just arisen, and–to use one of the most egregiously wrong and misleading phrases that ever dropped from the lips of man–the demand created the supply; it was quite the other way about. There was really no urgent demand for such things at the time; the current needs of the European world seem to have been fairly well served by coach and diligence in 1800, and, on the other hand, every administrator of intelligence in the Roman and Chinese empires must have felt an urgent need for more rapid methods of transit than those at his disposal. Nor was the development of the steam locomotive the result of any sudden discovery of steam. Steam, and something of the mechanical possibilities of steam, had been known for two thousand years; it had been used for pumping water, opening doors, and working toys, before the Christian era. It may be urged that this advance was the outcome of that new and more systematic handling of knowledge initiated by Lord Bacon and sustained by the Royal Society; but this does not appear to have been the case, though no doubt the new habits of mind that spread outward from that centre played their part. The men whose names are cardinal in the history of this development invented, for the most part, in a quite empirical way, and Trevithick's engine was running along its rails and Evan's boat was walloping up the Hudson a quarter of a century before Carnot expounded his general proposition. There were no such deductions from principles to application as occur in the story of electricity to justify our attribution of the steam-engine to the scientific impulse. Nor does this particular invention seem to have been directly due to the new possibilities of reducing, shaping, and casting iron, afforded by the substitution of coal for wood in iron-works; through the greater temperature afforded by a coal fire. In China coal has been used in the reduction of iron for many centuries. No doubt these new facilities did greatly help the steam-engine in its invasion of the field of common life, but quite certainly they were not sufficient to set it going. It was, indeed, not one cause, but a very complex and unprecedented series of causes, that set the steam locomotive going. It was indirectly, and in another way, that the introduction of coal became the decisive factor. One peculiar condition of its production in England seems to have supplied just one ingredient that had been missing for two thousand years in the group of conditions that were necessary before the steam locomotive could appear.

This missing ingredient was a demand for some comparatively simple, profitable machine, upon which the elementary principles of steam utilisation could be worked out. If one studies Stephenson's "Rocket" in detail, as one realises its profound complexity, one begins to understand how impossible it would have been for that structure to have come into existence *de novo,* however urgently the world had need of it. But it happened that the coal needed to replace the dwindling forests of this small and exceptionally rain-saturated country occurs in low, hollow basins overlying clay, and not, as in China and the Alleghanies for example, on high-lying outcrops, that can be worked as chalk is worked in England. From this fact it followed that some quite unprecedented pumping appliances became necessary, and the thoughts of practical men were turned thereby to the long-neglected possibilities of steam. Wind was extremely inconvenient for the purpose of pumping, because in these latitudes it is inconstant: it was costly, too, because at any time the labourers might be obliged to sit at the pit's mouth for weeks together, whistling for a gale or waiting for the water to be got under again. But steam had already been used for pumping upon one or two estates in England–rather as a toy than in earnest–before the middle of the seventeenth century, and the attempt to employ it was so obvious as to be practically unavoidable.* The water trickling into the coal-measures** acted, therefore, like water trickling upon chemicals that have long been mixed together dry and inert. Immediately the latent reactions were set going. Savery, Newcomen, a host of other workers, culminating in Watt, working always by steps that were at least so nearly obvious as to give rise again and again to simultaneous discoveries, changed this toy of steam into a real, a commercial thing, developed a trade in pumping-engines, created foundries and a new art of engineering, and almost unconscious of what they were doing, made the steam locomotive a well-nigh unavoidable consequence. At last, after a century of improvement on pumping-engines, there remained nothing but the very obvious stage of getting the engine that had been developed on wheels and out upon the ways of the world.

Ever and again during the eighteenth century an engine would be put upon the roads and pronounced a failure–one monstrous Palæoferric creature was visible on a French highroad as early as 1769–but by the dawn of the nineteenth century the problem had very nearly got itself solved. By 1804 Trevithick had a steam locomotive indisputably in motion and almost financially possible, and from his hands it puffed its way, slowly at first, and then, under Stephenson, faster and faster, to a transitory empire over the earth. It was a steam locomotive–but for all that it was primarily *a steam-engine for pumping* adapted to a new end; it was a steam-engine whose ancestral stage had developed under conditions that were by no means exacting in the matter of weight. And from that fact followed a consequence that has hampered railway travel and transport very greatly, and that is tolerated nowadays only through a belief in its practical necessity. The steam locomotive was all too huge and heavy for the highroad–it had to be put upon rails. And so clearly linked are steam-engines and railways in our minds that, in common language now, the latter implies the former. But indeed it is the result of accidental impediments, of avoidable difficulties that we travel to-day on rails.

Railway travelling is at best a compromise. The quite conceivable ideal of locomotive convenience, so far as travellers are

*It might have been used in the same way in Italy in the first century, had not the grandiose taste for aqueducts prevailed.

**And also into the Cornwall mines, be it noted.

concerned, is surely a highly mobile conveyance capable of travelling easily and swiftly to any desired point, traversing, at a reasonably controlled pace, the ordinary roads and streets, and having access for higher rates of speed and long-distance travelling to specialised ways restricted to swift traffic, and possibly furnished with guide-rails. For the collection and delivery of all sorts of perishable goods also the same system is obviously altogether superior to the existing methods. Moreover, such a system would admit of that secular progress in engines and vehicles that the stereotyped conditions of the railway have almost completely arrested, because it would allow almost any new pattern to be put at once upon the ways without interference with the established traffic. Had such an ideal been kept in view from the first the traveller would now be able to get through his long-distance journeys at a pace of from seventy miles or more an hour without changing, and without any of the trouble, waiting, expense, and delay that arises between the household or hotel and the actual rail. It was an ideal that must have been at least possible to an intelligent person fifty years ago, and, had it been resolutely pursued, the world, instead of fumbling from compromise to compromise as it always has done and as it will do very probably for many centuries yet, might have been provided to-day, not only with an infinitely more practicable method of communication, but with one capable of a steady and continual evolution from year to year.

But there was a more obvious path of development and one immediately cheaper, and along that path went short-sighted Nineteenth Century Progress, quite heedless of the possibility of ending in a *cul-de-sac*. The first locomotives, apart from the heavy tradition of their ancestry, were, like all experimental machinery, needlessly clumsy and heavy, and their inventors, being men of insufficient faith, instead of working for lightness and smoothness of motion, took the easier course of placing them upon the tramways that were already in existence—chiefly for the transit of heavy goods over soft roads. And from that followed a very interesting and curious result.

These tram-lines very naturally had exactly the width of an ordinary cart, a width prescribed by the strength of one horse. Few people saw in the locomotive anything but a cheap substitute for horseflesh, or found anything incongruous in letting the dimensions of a horse determine the dimensions of an engine. It mattered nothing that from the first the passenger was ridiculously cramped, hampered, and crowded in the carriage. He had always been cramped in a coach, and it would have seemed "Utopian"–a very dreadful thing indeed to our grandparents–to propose travel without cramping. By mere inertia the horse-cart gauge, the 4 ft. 8½ in. gauge, *nemine contradicente,* established itself in the world, and now everywhere the train is dwarfed to a scale that limits alike its comfort, power, and speed. Before every engine, as it were, trots the ghost of a superseded horse, refuses most resolutely to trot faster than fifty miles an hour, and shies and threatens catastrophe at every point and curve. That fifty miles an hour, most authorities are agreed, is the limit of our speed for land travel, so far as existing conditions go.* Only a revolutionary reconstruction of the railways or the development of some new competing method of land travel can carry us beyond that.

People of to-day take the railways for granted as they take sea and sky; they were born in a railway world, and they expect to die in one. But if only they will strip from their eyes the most blinding of all influences, acquiescence in the familiar, they will see clearly enough that this vast and elaborate railway-system of ours, by which the whole world is linked together, is really only a vast system of trains of horse-waggons and coaches drawn along rails by pumping-engines upon wheels. Is that, in spite of its present vast extension, likely to remain the predominant method of land locomotion–even for so short a period as the next hundred years?

Now, so much capital is represented by the existing type of railways, and they have so firm an establishment in the acquiescence of men, that it is very doubtful if the railways will ever attempt any very fundamental change in the direction of greater speed or facility, unless they are first exposed to the pressure of our second alternative, competition, and we may very well go on to inquire how long will it be before that second alternative comes into operation–if ever it is to do so.

Let us consider what other possibilities seem to offer themselves. Let us revert to the ideal we have already laid down, and consider what hopes and obstacles to its attainment there seem

*It might be worse. If the biggest horses had been Shetland ponies, we should be travelling now in railway-carriages to hold two each side at a maximum speed of perhaps twenty miles an hour. There is hardly any reason, beyond this tradition of the horse, why the railway-carriage should not be even nine or ten feet wide, the width, that is, of the smallest room in which people can live in comfort, hung on such springs and wheels as would effectually destroy all vibration, and furnished with all the equipment of comfortable chambers.

to be. The abounding presence of numerous experimental motors to-day is so stimulating to the imagination, there are so many stimulated persons at work upon them, that it is difficult to believe the obvious impossibility of most of them–their convulsiveness, clumsiness, and, in many cases, exasperating trail of stench will not be rapidly fined away.* I do not think that

*Explosives as a motive power were first attempted by Huyghens and one or two others in the seventeenth century, and, just as with the turbine type of apparatus, it was probably the impetus given to the development of steam by the convenient collocation of coal and water and the need of an engine, that arrested the advance of this parallel inquiry until our own time. Explosive engines, in which gas and petroleum are employed, are now abundant, but for all that we can regard the explosive engine as still in its experimental stages. So far, research in explosives has been directed chiefly to the possibilities of higher and still higher explosives for use in war, the neglect of the mechanical application of this class of substance being largely due to the fact, that chemists are not as a rule engineers, nor engineers chemists. But an easily portable substance, the decomposition of which would evolve energy, or–what is, from the practical point of view, much the same thing–an easily portable substance, which could be decomposed electrically by wind or water power, and which would then recombine and supply force, either in intermittent thrusts at a piston, or as an electric current, would be infinitely more convenient for all locomotive purposes than the cumbersome bunkers and boilers required by steam. The presumption is altogether in favour of the possibility of such substances. Their advent will be the beginning of the end for steam traction on land and of the steamship at sea: the end indeed of the Age of Coal and Steam. And even with regard to steam there may be a curious change of method before the end. It is beginning to appear that, after all, the piston and cylinder type of engine is, for locomotive purposes–on water at least, if not on land–by no means the most perfect. Another, and fundamentally different type, the turbine type, in which the impulse of the steam spins a wheel instead of shoving a piston, would appear to be altogether better than the adapted pumping-engine, at anyrate, for the purposes of steam navigation. Hero, of Alexandria, describes an elementary form of such an engine, and the early experimenters of the seventeenth century tried and abandoned the rotary principle. It was not adapted to pumping, and pumping was the only application that then offered sufficient immediate encouragement to persistence. The thing marked time for quite two centuries and a half, therefore, while the piston-engines perfected themselves; and only in the eighties did the requirements of the dynamo-electric machine open a "practicable" way of advance. The motors of the dynamo-electric machine in the nineteenth century, in fact, played exactly the *rôle* of the pumping-engine in the eighteenth, and by 1894 so many difficulties of detail had been settled, that a syndicate of capitalists and scientific men could face the construction of an experimental ship. This ship, the *Turbinia,* after a considerable amount of trial and modification, attained the unprecedented speed of 34½ knots an hour, and His Majesty's navy has possessed, in the *Turbinia's* younger and greater sister, the *Viper,* now unhappily lost, a torpedo-destroyer capable of 41 miles an hour. There can be little doubt that the sea speeds of 50 and even 60 miles an hour will be attained within the next few years. But I do not think that these developments will do more than delay the advent of the "explosive" or "storage of force" engine.

it is asking too much of the reader's faith in progress to assume that so far as a light, powerful engine goes, comparatively noiseless, smooth-running, not obnoxious to sensitive nostrils, and altogether suitable for highroad traffic, the problem will very speedily be solved. And upon that assumption, in what direction are these new motor vehicles likely to develop? how will they react upon the railways? and where finally will they take us?

At present they seem to promise developments upon three distinct and definite lines.

There will, first of all, be the motor truck for heavy traffic. Already such trucks are in evidence distributing goods and parcels of various sorts. And sooner or later, no doubt, the numerous advantages of such an arrangement will lead to the organisation of large carrier companies, using such motor trucks to carry goods in bulk or parcels on the highroads. Such companies will be in an exceptionally favourable position to organise storage and repair for the motors of the general public on profitable terms, and possibly to co-operate in various ways with the manufactures of special types of motor machines.

In the next place, and parallel with the motor truck, there will develop the hired or privately owned motor carriage. This, for all except the longest journeys, will add a fine sense of personal independence to all the small conveniences of first-class railway travel. It will be capable of a day's journey of three hundred miles or more, long before the developments to be presently foreshadowed arrive. One will change nothing–unless it is the driver–from stage to stage. One will be free to dine where one chooses, hurry when one chooses, travel asleep or awake, stop and pick flowers, turn over in bed of a morning and tell the carriage to wait–unless, which is highly probable, one sleeps aboard.* . . .

*The historian of the future, writing about the nineteenth century, will, I sometimes fancy, find a new meaning in a familiar phrase. It is the custom to call this the most "Democratic" age the world has ever seen, and most of us are beguiled by the etymological contrast, and the memory of certain legislative revolutions, to oppose one form of stupidity prevailing to another, and to fancy we mean the opposite to an "Aristocratic" period. But indeed we do not. So far as that political point goes, the Chinaman has always been infinitely more democratic than the European. But the world, by a series of gradations into error, has come to use "Democratic" as a substitute for "Wholesale," and as an opposite to "Individual," without realising the shifted application at all. Thereby old "Aristocracy," the organisation of society for the glory and preservation of the Select Dull, gets to a flavour even of freedom. When the historian of the future speaks of the past century as a Democratic century, he will have in mind, more than anything else, the unprecedented fact that we seemed to do

And thirdly there will be the motor omnibus, attacking or developing out of the horse omnibus companies and the suburban lines. All this seems fairly safe prophesying.

And these things, which are quite obviously coming even now, will be working out their many structural problems when the next phase in their development begins. The motor omnibus companies competing against the suburban railways will find themselves hampered in the speed of their longer runs by the slower horse traffic on their routes, and they will attempt to secure, and, it may be, after tough legislative struggles, will secure the power to form private roads of a new sort, upon which their vehicles will be free to travel up to the limit of their very highest possible speed. It is along the line of such private tracks and roads that the forces of change will certainly tend to travel, and along which I am absolutely convinced they will travel. This segregation of motor traffic is probably a matter that may begin even in the present decade.

Once this process of segregation from the highroad of the horse and pedestrian sets in, it will probably go on rapidly. It may spread out from short omnibus routes, much as the London Metropolitan Railway system has spread. The motor carrier companies, competing in speed of delivery with the quickened railways, will conceivably co-operate with the long-distance omnibus and the hired carriage companies in the formation of trunk-lines. Almost insensibly, certain highly profitable longer routes will be joined up–the London to Brighton, for example, in England. And the quiet English citizen will, no doubt, while these things are still quite exceptional and experimental in his lagging land, read one day with surprise in the violently illustrated popular magazines of 1910, that there are now so many thousand miles of these roads already established in America and Germany and elsewhere. And thereupon, after some patriotic meditations, he may pull himself together.

We may even hazard some details about these special roads.

everything in heaps–we read in epidemics; clothed ourselves, all over the world, in identical fashions; built and furnished our houses in stereo designs; and travelled–that naturally most individual proceeding–in bales. To make the railway-train a perfect symbol of our times, it should be presented as uncomfortably full in the third class–a few passengers standing–and everybody reading the current number either of the *Daily Mail, Pearson's Weekly, Answers, Tit Bits,* or whatever Greatest Novel of the Century happened to be going. . . . But, as I hope to make clearer in my later papers, this "Democracy," or Wholesale method of living, like the railways, is transient–a first makeshift development of a great and finally (to me at least) quite hopeful social reorganisation.

For example, they will be very different from macadamised roads; they will be used only by soft-tired conveyances; the battering horse-shoes, the perpetual filth of horse-traffic, and the clumsy wheels of laden carts will never wear them. It may be that they will have a surface like that of some cycle-racing tracks, though since they will be open to wind and weather, it is perhaps more probable they will be made of very good asphalt sloped to drain, and still more probable that they will be of some quite new substance altogether–whether hard or resilient is beyond my foretelling. They will have to be very wide–they will be just as wide as the courage of their promoters goes–and if the first made are too narrow there will be no question of gauge to limit the later ones. Their traffic in opposite directions will probably be strictly separated, and it will no doubt habitually disregard complicated and fussy regulations imposed under the initiative of the Railway Interest by such official bodies as the Board of Trade. The promoters will doubtless take a hint from suburban railway-traffic and from the current difficulty of the Metropolitan police, and where their ways branch the streams of traffic will not cross at a level but by bridges. It is easily conceivable that once these tracks are in existence, cyclists and motors other than those of the constructing companies will be able to make use of them. And, moreover, once they exist it will be possible to experiment with vehicles of a size and power quite beyond the dimensions prescribed by our ordinary roads–roads whose width has been entirely determined by the size of a cart a horse can pull.*

Countless modifying influences will, of course, come into operation. For example, it has been assumed, perhaps rashly, that the railway influence will certainly remain jealous and hostile to these growths: that what may be called the "Bicycle Ticket Policy" will be pursued throughout. Assuredly there will be fights of a very complicated sort at first, but once one of these specialised lines is in operation, it may be that some at least of the railway-companies will hasten to replace their flanged rolling-stock by carriages with rubber tyres, remove their rails, broaden their cuttings and embankments, raise their bridges, and take to the new ways of traffic. Or they may find it answer to cut fares, widen their gauges, reduce their gradients, modify their points and curves, and woo the passenger back with carriages beautifully hung and sumptuously furnished, and all the convenience and

*So we begin to see the possibility of laying that phantom horse that haunts the railways to this day so disastrously.

luxury of a club. Few people would mind being an hour or so longer going to Paris from London, if the railway travelling was neither rackety, cramped, nor tedious. One could be patient enough if one was neither being jarred, deafened, cut into slices by draughts, and continually more densely caked in a filthy dust of coal; if one could write smoothly and easily at a steady table, read papers, have one's hair cut, and dine in comfort*–none of which things are possible at present, and none of which require any new inventions, any revolutionary contrivances, or indeed anything but an intelligent application of existing resources and known principles. Our rage for fast trains, so far as long-distance travel is concerned, is largely a passion to end the extreme discomfort involved. It is in the daily journey, on the suburban train, that daily tax of time, that speed is in itself so eminently desirable, and it is just here that the conditions of railway travel most hopelessly fail. It must always be remembered that the railway-train, as against the motor, has the advantage that its wholesale traction reduces the prime cost by demanding only one engine for a great number of coaches. This will not serve the first-class long-distance passenger, but it may the third. Against that economy one must balance the necessary delay of a relatively infrequent service, which latter item becomes relatively greater and greater in proportion to the former, the briefer the journey to be made.

*A correspondent, Mr. Rudolf Cyrian, writes to correct me here, and I cannot do better, I think, than thank him and quote what he says. "It is hardly right to state that fifty miles an hour 'is the limit of our speed for land travel, so far as existing conditions go.' As far as English traffic is concerned, the statement is approximately correct. In the United States, however, there are several trains running now which average over considerable distances more than sixty miles and hour, stoppages included, nor is there much reason why this should not be considerably increased. What especially hampers the development of railways in England–as compared with other countries–is the fact that the rolling-stock templet is too small. Hence carriages in England have to be narrower and lower than carriages in the United States, although both run on the same standard gauge (4 feet 8½ inches). The result is that several things which you describe as not possible at present, such as to 'write smoothly and easily at a steady table, read papers, have one's hair cut, and dine in comfort,' are not only feasible, but actually attained on some of the good American trains. For instance, on the *present* Empire State Express, running between New York and Buffalo, or on the *present* Pennsylvania, Limited, running between New York and Chicago, and on others. With the Pennsylvania, Limited, travel stenographers and typewriters, whose services are placed at the disposal of passengers free of charge. But the train on which there is the least vibration of any is probably the new Empire State Express, and on this it is certainly possible to write smoothly and easily at a steady table."

And it may be that many railways, which are neither capable of modification into suburban motor tracks, nor of development into luxurious through routes, will find, in spite of the loss of many elements of their old activity, that there is still a profit to be made from a certain section of the heavy goods traffic, and from cheap excursions. These are forms of work for which railways seem to be particularly adapted, and which the diversion of a great portion of their passenger traffic would enable them to conduct even more efficiently. It is difficult to imagine, for example, how any sort of road-car organisation could beat the railways at the business of distributing coal and timber and similar goods, which are taken in bulk directly from the pit or wharf to local centres of distribution.

It must always be remembered that at the worst the defeat of such a great organisation as the railway-system does not involve its disappearance until a long period has elapsed. It means at first no more than a period of modification and differentiation. Before extinction can happen a certain amount of wealth in railway property must absolutely disappear. Though under the stress of successful competition that the capital value of the railways may conceivably fall, and continue to fall, towards the marine store prices, fares and freights pursue the sweated working expenses to the vanishing-point, and the land occupied sink to the level of not very eligible building-sites: yet the railways will, nevertheless, continue in operation until these downward limits are positively attained.

An imagination prone to the picturesque insists at this stage upon a vision of the latter days of one of the less happily situated lines. Along a weedy embankment there pants and clangs a patched and tarnished engine, its paint blistered, its parts leprously dull. It is driven by an aged and sweated driver, and the burning garbage of its furnace distills a choking reek into the air. A huge train of urban dust trucks bangs and clatters behind it, *en route* to that sequestered dumping-ground where rubbish is burnt to some industrial end. But that is a lapse into the merely just possible, and at most a local tragedy. Almost certainly the existing lines of railway will develop and differentiate, some in one direction and some in another, according to the nature of the pressure upon them. Almost all will probably be still in existence and in divers ways busy, spite of the swarming new highways I have ventured to foreshadow, a hundred years from now.

In fact, we have to contemplate, not so much a supersession of the railways as a modification and specialisation of them in various directions, and the enormous development beside them of competing and supplementary methods. And step by step with these developments will come a very considerable acceleration of the ferry traffic of the narrow seas through such improvements as the introduction of turbine engines. So far as the high-road and the longer journeys go this is the extent of our prophecy.*

But in the discussion of all questions of land locomotion one must come at last to the knots of the network, to the central portions of the towns, the dense, vast towns of our time, with their high ground-values and their narrow, already almost impassable, streets. I hope at a later stage to give some reasons for anticipating that the centripetal pressure of the congested towns of our epoch may ultimately be very greatly relieved, but for the next few decades at least the usage of existing conditions will prevail, and in every town there is a certain nucleus of offices, hotels, and shops upon which the centrifugal forces I anticipate will certainly not operate. At present the streets of many larger towns, and especially of such old-established towns as London, whose central portions have the narrowest arteries, present a quite unprecedented state of congestion. When the Green of some future *History of the English People* comes to review our times, he will, from his standpoint of comfort and convenience, find the present streets of London quite or even more incredibly unpleasant than are the filthy kennels, the mud-holes and darkness of the streets of the seventeenth century to our enlightened minds. He will echo our question, "Why *did* people stand it?" He will be struck first of all by the omnipresence of mud, filthy mud, churned up by hoofs and wheels under the inclement skies, and perpetually defiled and added to by innumerable horses. Imagine his description of a young lady crossing the road at the Marble Arch in London, on a wet November afternoon, "breathless, foul-footed, splashed by a passing hansom from head to foot, happy that she has reached the further pavement alive at the mere cost of her ruined clothes." . . . "Just where the bicycle might have served its most useful purpose," he will write, "in affording a healthy daily ride to the innumerable clerks and such-like sedentary toilers of the central region, it was rendered

*Since this appeared in the *Fortnightly Review* I have had the pleasure of reading 'Twentieth Century Inventions,' by Mr. George Sutherland, and I find very much else of interest bearing on these questions–the happy suggestion (for the ferry transits, at anyrate) of a rail along the sea bottom, which would serve as a guide to swift submarine vessels, out of reach of all that superficial "motion" that is so distressing, and of all possibilities of collision.

impossible by the danger of side-slip in this vast ferocious traffic." And, indeed, to my mind at least, this last is the crowning absurdity of the present state of affairs, that the clerk and the shop hand, classes of people positively starved of exercise, should be obliged to spend yearly the price of a bicycle upon a season-ticket, because of the quite unendurable inconvenience and danger of urban cycling.

Now, in what direction will matters move? The first and most obvious thing to do, the thing that in many cases is being attempted and in a futile, insufficient way getting itself done, the thing that I do not for one moment regard as the final remedy, is the remedy of the architect and builder–profitable enough to them, anyhow–to widen the streets and to cut "new arteries." Now, every new artery means a series of new whirlpools of traffic, such as the pensive Londoner may study for himself at the intersection of Shaftesbury Avenue with Oxford Street, and unless colossal–or inconveniently steep–crossing-bridges are made, the wider the affluent arteries the more terrible the battle of the traffic. Imagine Regent's Circus on the scale of the Place de la Concorde. And there is the value of the ground to consider; with every increment of width the value of the dwindling remainder in the meshes of the network of roads will rise, until to pave the widened streets with gold will be a mere trifling addition to the cost of their "improvement."

There is, however, quite another direction in which the congestion may find relief, and that is in the "regulation" of the traffic. This has already begun in London in an attack on the crawling cab and in the new bye-laws of the London County Council, whereby certain specified forms of heavy traffic are prohibited the use of the streets between ten and seven. These things may be the first beginning of a process of restriction that may go far. Many people living at the present time, who have grown up amidst the exceptional and possibly very transient characteristics of this time, will be disposed to regard the traffic in the streets of our great cities as a part of the natural order of things, and as unavoidable as the throng upon the pavement. But indeed the presence of all the chief constituents of this vehicular torrent–the cabs and hansoms, the vans, the omnibuses–everything, indeed, except the few private carriages–are as novel, as distinctively things of the nineteenth century, as the railway-train and the needle telegraph. The streets of the great towns of antiquity, the streets of the great towns of the East, the streets of all the mediæval towns, were not intended for any sort of wheeled traffic at all–were designed primarily and chiefly for pedestrians. So it would be, I suppose, in anyone's ideal city. Surely Town, in theory at least, is a place one walks about as one walks about a house and garden, dressed with a certain ceremonious elaboration, safe from mud and the hardship and defilement of foul weather, buying, meeting, dining, studying, carousing, seeing the play. It is the growth in size of the city that has necessitated the growth of this coarser traffic that has made "Town" at last so utterly detestable.

But if one reflects, it becomes clear that, save for the vans of goods, this moving tide of wheeled masses is still essentially a stream of urban pedestrians, pedestrians who, by reason of the distances they have to go, have had to jump on 'buses and take cabs–in a word, to bring in the highroad to their aid. And the vehicular traffic of the street is essentially the highroad traffic very roughly adapted to the new needs. The cab is a simple development of the carriage, the omnibus of the coach, and the supplementary traffic of the underground and electric railways is a by no means brilliantly imagined adaptation of the long-route railway. These are all still new things, experimental to the highest degree, changing and bound to change much more, in the period of specialisation that is now beginning.

Now, the first most probable development is a change in the omnibus and the omnibus railway. A point quite as important with these means of transit as actual speed of movement is frequency: time is wasted abundantly and most vexatiously at present in waiting and in accommodating one's arrangements to infrequent times of call and departure. *The more frequent a local service, the more it comes to be relied upon.* Another point–and one in which the omnibus has a great advantage over the railway–is that it should be possible to get on and off at any point, or at as many points on the route as possible. But this means a high proportion of stoppages, and this is destructive to speed. There is, however, one conceivable means of transit that is not simply frequent but continuous, that may be joined or left at any point without a stoppage, that could be adapted to many existing streets at the level or quite easily sunken in tunnels, or elevated above the street level,* and that means of transit is the moving platform, whose possibilities have been exhibited to all the world in a sort of mean caricature at the Paris Exhibition. Let us

*To the level of such upper storey pavements as Sir F. Bramwell has proposed for the new Holborn to Strand Street, for example.

imagine the inner circle of the district railway adapted to this conception. I will presume that the Parisian "rolling platform" is familiar to the reader. The district railway tunnel is, I imagine, about twenty-four feet wide. If we suppose the space given to six platforms of three feet wide and one (the most rapid) of six feet, and if we suppose each platform to be going four miles an hour faster than its slower fellow (a velocity the Paris experiment has shown to be perfectly comfortable and safe), we should have the upper platform running round the circle at a pace of twenty-eight miles an hour. If, further, we adopt an ingenious suggestion of Professor Perry's, and imagine the descent to the line made down a very slowly rotating staircase at the centre of a big rotating wheel-shaped platform, against a portion of whose rim the slowest platform runs in a curve, one could very easily add a speed of six or eight miles an hour more, and to that the man in a hurry would be able to add his own four miles an hour by walking in the direction of motion. If the reader is a traveller, and if he will imagine that black and sulphurous tunnel, swept and garnished, lit and sweet, with a train much faster than the existing underground trains perpetually ready to go off with him and never crowded–if he will further imagine this train a platform set with comfortable seats and neat bookstalls and so forth, he will get an inkling in just one detail of what he perhaps misses by living now instead of thirty or forty years ahead.

I have supposed the replacement to occur in the case of the London Inner Circle Railway, because there the necessary tunnel already exists to help the imagination of the English reader, but that the specific replacement will occur is rendered improbable by the fact that the circle is for much of its circumference entangled with other lines of communication–the North-Western Railway, for example. As a matter of fact, as the American reader at least will promptly see, the much more practicable thing is that upper footpath, with these moving platforms beside it, running out over the street after the manner of the viaduct of an elevated railroad. But in some cases, at anyrate, the demonstrated cheapness and practicability of tunnels at a considerable depth will come into play.

Will this diversion of the vast omnibus traffic of to-day into the air and underground, together with the segregation of van traffic to specific routes and times, be the only change in the streets of the new century? It may be a shock, perhaps, to some minds, but I must confess I do not see what is to prevent the process of elimination that is beginning now with the heavy vans spreading until it covers all horse-traffic, and with the disappearance of horse-hoofs and the necessary filth of horses, the road surface may be made a very different thing from what it is at present, better drained and admirably adapted for the soft-tired hackney vehicles and the torrent of cyclists. Moreover, there will be little to prevent a widening of the existing side-walks, and the protection of the passengers from rain and hot sun by awnings, or such arcades as distinguish Turin, or Sir F. Bramwell's upper footpaths on the model of the Chester rows. Moreover, there is no reason but the existing filth why the roadways should not have translucent *velaria* to pull over in bright sunshine and wet weather. It would probably need less labour to manipulate such contrivances than is required at present for the constant conflict with slush and dust. Now, of course, we tolerate the rain, because it facilitates a sort of cleaning process. . . .

Enough of this present speculation. I have indicated now the general lines of the roads and streets and ways and underways of the Twentieth Century. But at present they stand vacant in our prophecy, not only awaiting the human interests–the characters and occupations, and clothing of the throng of our children and our children's children that flows along them, but also the decorations our children's children's taste will dictate, the advertisements their eyes will tolerate, the shops in which they will buy. To all that we shall finally come, and even in the next chapter I hope it will be made more evident how conveniently these later and more intimate matters follow, instead of preceding, these present mechanical considerations. And of the beliefs and hopes, the thought and language, the further prospects of this multitude as yet unborn–of these things also we shall make at last certain hazardous guesses. But at first I would submit to those who may find the "machinery in motion" excessive in this chapter, we must have the background and fittings–the scene before the play.*

*I have said nothing in this chapter, devoted to locomotion, of the coming invention of flying. This is from no disbelief in its final practicability, nor from any disregard of the new influences it will bring to bear upon mankind. But I do not think it at all probable that aeronautics will ever come into play as a serious modification of transport and communication–the main question here under consideration. Man is not, for example, an albatross, but a land biped, with a considerable disposition towards being made sick and giddy by unusual motions, and however he soars he must come to earth to live. We must build our picture of the future from the ground upward; of flying–in its place.

This chapter has been very ably criticised in many of its details in the reviews of the first edition, but I do not think anything has been said to undermine the general proposition I have advanced nor to affect the conclusions drawn in the following chapter. I have ignored the need of guide-rails for specialised high-speed roads such as are described on [pp. 10–11], and the possibility (which my friend Mr. Joseph Conrad has suggested to me) of sliding cars along practically frictionless rails.

II. THE PROBABLE DIFFUSION OF GREAT CITIES

NOW, the velocity at which a man and his belongings may pass about the earth is in itself a very trivial matter indeed, but it involves certain other matters not at all trivial, standing, indeed, in an almost fundamental relation to human society. It will be the business of this chapter to discuss the relation between the social order and the available means of transit, and to attempt to deduce from the principles elucidated the coming phases in that extraordinary expansion, shifting and internal redistribution of population that has been so conspicuous during the last hundred years.

Let us consider the broad features of the redistribution of the population that has characterised the nineteenth century. It may be summarised as an unusual growth of great cities and a slight tendency to depopulation in the country. The growth of the great cities is the essential phenomenon. These aggregates having populations of from eight hundred thousand upward to four and five millions, are certainly, so far as the world outside the limits of the Chinese empire goes, entirely an unprecedented thing. Never before, outside the valleys of the three great Chinese rivers, has any city–with the exception of Rome and perhaps (but very doubtfully) of Babylon–certainly had more than a million inhabitants, and it is at least permissible to doubt whether the population of Rome, in spite of its exacting a tribute of sea-borne food from the whole of the Mediterranean basin, exceeded a million for any great length of time.* But there are

*It is true that many scholars estimate a high-water mark for the Roman population in excess of two millions; and one daring authority, by throwing

now ten town aggregates having a population of over a million, nearly twenty that bid fair to reach that limit in the next decade, and a great number at or approaching a quarter of a million. We call these towns and cities, but, indeed, they are of a different order of things to the towns and cities of the eighteenth-century world.

Concurrently with the aggregation of people about this new sort of centre, there has been, it is alleged, a depletion of the country villages and small townships. But, so far as the counting of heads goes, this depletion is not nearly so marked as the growth of the great towns. Relatively, however, it is striking enough.

Now, is this growth of large towns really, as one may allege, a result of the development of railways in the world, or is it simply a change in human circumstances that happens to have arisen at the same time? It needs only a very general review of the conditions of the distribution of population to realise that the former is probably the true answer.

It will be convenient to make the issue part of a more general proposition, namely, that *the general distribution of population in a country must always be directly dependent on transport facilities.* To illustrate this point roughly we may build up an imaginary simple community by considering its needs. Over an arable countryside, for example, inhabited by a people who had attained to a level of agricultural civilisation in which war was no longer constantly imminent, the population would be diffused primarily by families and groups in farmsteads. It might, if it were a very simple population, be almost all so distributed. But even the simplest

out suburbs *ad libitum* into the Campagna, suburbs of which no trace remains, has raised the two to ten. The Colosseum could, no doubt, seat over 80,000 spectators; the circuit of the bench frontage of the Circus Maximus was very nearly a mile in length, and the Romans of Imperial times certainly used ten times as much water as the modern Romans. But, on the other hand, habits change, and Rome as it is defined by lines drawn at the times of its greatest ascendancy–the city, that is, enclosed by the walls of Aurelian and including all the *regiones* of Augustus, an enclosure from which there could have been no reason for excluding half or more of its population–could have scarcely contained a million. It would have packed very comfortably within the circle of the Grands Boulevards of Paris–the Paris, that is, of Louis XIV., with a population of 560,000; and the Rome of to-day, were the houses that spread so densely over the once vacant Campus Martius distributed in the now deserted spaces in the south and east, and the Vatican suburb replaced within the ancient walls, would quite fill the ancient limits, in spite of the fact that the population is under 500,000. But these are incidental doubts on a very authoritative opinion, and, whatever their value, they do not greatly affect the significance of these new great cities, which have arisen all over the world, as if by the operation of a natural law, as the railways have developed.

agriculturists find a certain convenience in trade. Certain definite points would be convenient for such local trade and intercourse as the people found desirable, and here it is that there would arise the germ of a town. At first it might be no more than an appointed meeting-place, a market square, but an inn and a blacksmith would inevitably follow, an altar, perhaps, and, if these people had writing, even some sort of school. It would have to be where water was found, and it would have to be generally convenient of access to its attendant farmers.

Now, if this meeting-place was more than a certain distance from any particular farm, it would be inconvenient for that farmer to get himself and his produce there and back, and to do his business in a comfortable daylight. He would not be able to come and, instead, he would either have to go to some other nearer centre to trade and gossip with his neighbours or, failing this, not go at all. Evidently, then, there would be a maximum distance between such places. This distance in England, where traffic has been mainly horse-traffic for many centuries, seems to have worked out, according to the gradients and so forth, at from eight to fifteen miles, and at such distances do we find the country towns, while the horseless man, the serf, and the labourer and labouring wench have marked their narrow limits in the distribution of the intervening villages. If by chance these gathering-places have arisen at points much closer than this maximum, they have come into competition, and one has finally got the better of the other, so that in England the distribution is often singularly uniform. Agricultural districts have their towns at about eight miles, and where grazing takes the place of the plough, the town distances increase to fifteen.* And so it is, entirely as a multiple of horse and foot strides, that all the villages and towns of the world's country-side have been plotted out**

A third and almost final, factor determining town distribution in a world without railways, would be the sea-port and the navigable river. Ports would grow into dimensions dependent on the population of the conveniently accessible coasts (or river-banks),

*It will be plain that such towns must have clearly defined limits of population, *dependent finally on the minimum yearly produce of the district they control.* If ever they rise above that limit the natural checks of famine, and of pestilence following enfeeblement, will come into operation, and they will always be kept near this limit by the natural tendency of humanity to increase. The limit would rise with increasing public intelligence, and the organisation of the towns would become more definite.

**I owe the fertilising suggestion of this general principle to a paper by Grant Allen that I read long ago in *Longman's Magazine.*

and on the quality and quantity of their products, and near these ports, as the conveniences of civilisation increased, would appear handicraft towns–the largest possible towns of a foot-and-horse civilisation–with industries of such a nature as the produce of their coasts required.

It was always in connection with a port or navigable river that the greater towns of the pre-railway periods arose, a day's journey away from the coast when sea attack was probable, and shifting to the coast itself when that ceased to threaten. Such sea-trading handicraft towns as Bruges, Venice, Corinth, or London were the largest towns of the vanishing order of things. Very rarely, except in China, did they clamber above a quarter of a million inhabitants, even though to some of them there was presently added court and camp. In China, however, a gigantic river and canal system, laced across plains of extraordinary fertility, has permitted the growth of several city aggregates with populations exceeding a million, and in the case of the Hankow trinity of cities exceeding five million people.

In all these cases the position and the population limit was entirely determined by the accessibility of the town and the area it could dominate for the purposes of trade. And not only were the commercial or natural towns so determined, but the political centres were also finally chosen for strategic considerations, in a word–communications. And now, perhaps, the real significance of the previous paper, in which sea velocities of fifty miles an hour, and land travel at the rate of a hundred, and even cab and omnibus journeys of thirty or forty miles, were shown to be possible, becomes more apparent.

At the first sight it might appear as though the result of the new developments was simply to increase the number of giant cities in the world by rendering them possible in regions where they had hitherto been impossible–concentrating the trade of vast areas in a manner that had hitherto been entirely characteristic of navigable waters. It might seem as though the state of affairs in China, in which population has been concentrated about densely-congested "million-cities," with pauper masses, public charities, and a crowded struggle for existence, for many hundreds of years, was merely to be extended over the whole world. We have heard so much of the "problem of our great cities"; we have the impressive statistics of their growth; the belief in the inevitableness of yet denser and more multitudinous agglomerations in the future is so widely diffused, that at first sight it will be thought that no other motive than a wish to startle can dictate the proposition that not only will many of these railway-begotten "giant cities" reach their maximum in the commencing century, but that in all probability they, and not only they, but their water-born prototypes in the East also, are destined to such a process of dissection and diffusion as to amount almost to obliteration, so far, at least, as the blot on the map goes, within a measurable further space of years.

In advancing this proposition, the present writer is disagreeably aware that in this matter he has expressed views entirely opposed to those he now propounds; and in setting forth the following body of considerations he tells the story of his own disillusionment. At the outset he took for granted–and, very naturally, he wishes to imagine that a great number of other people do also take for granted–that the future of London, for example, is largely to be got as the answer to a sort of rule-of-three sum. If in one hundred years the population of London has been multiplied by seven, then in two hundred years—! And one proceeds to pack the answer in gigantic tenement houses, looming upon colossal roofed streets, provide it with moving ways (the only available transit appliances suited to such dense multitudes), and develop its manners and morals in accordance with the laws that will always prevail amidst over-crowded humanity so long as humanity endures. The picture of this swarming, concentrated humanity has some effective possibilities, but, unhappily, if, instead of that obvious rule-of-three sum, one resorts to an analysis of operating causes, its plausibility crumbles away, and it gives place to an altogether different forecast–a forecast, indeed, that is in almost violent contrast to the first anticipation. It is much more probable that these coming cities will not be, in the old sense, cities at all; they will present a new and entirely different phase of human distribution.

The determining factor in the appearance of great cities in the past, and, indeed, up to the present day, has been the meeting of two or more transit lines, the confluence of two or more streams of trade, and easy communication. The final limit to the size and importance of the great city has been the commercial "sphere of influence" commanded by that city, the capacity of the alluvial basin of its commerce, so to speak, the volume of its river of trade. About the meeting-point so determined the population so determined has grouped itself–and this is the point I overlooked in those previous vaticinations–in accordance with *laws that are also considerations of transit.*

The economic centre of the city is formed, of course, by the wharves and landing-places–and in the case of railway-fed cities by the termini–where passengers land and where goods are landed, stored, and distributed. Both the administrative and business community, traders, employers, clerks, and so forth, must be within a convenient access of this centre; and the families, servants, tradesmen, amusement purveyors dependent on these again must also come within a maximum distance. At a certain stage in town growth the pressure on the more central area would become too great for habitual family life there, and an office region would differentiate from an outer region of homes. Beyond these two zones, again, those whose connection with the great city was merely intermittent would constitute a system of suburban houses and areas. But the grouping of these, also, would be determined finally by the convenience of access to the dominant centre. That secondary centres, literary, social, political, or military, may arise about the initial trade centre, complicates the application but does not alter the principle here stated. They must all be within striking distance. The day of twenty-four hours is an inexorable human condition, and up to the present time all intercourse and business has been broken into spells of definite duration by intervening nights. Moreover, almost all effective intercourse has involved personal presence at the point where intercourse occurs. The possibility, therefore, of going and coming and doing that day's work has hitherto fixed the extreme limits to which a city could grow, and has exacted a compactness which has always been very undesirable and which is now for the first time in the world's history no longer imperative.

So far as we can judge without a close and uncongenial scrutiny of statistics, that daily journey, that has governed and still to a very considerable extent governs the growth of cities, has had, and probably always will have, a maximum limit of two hours, one hour each way from sleeping-place to council chamber, counter, work-room, or office stool. And taking this assumption as sound, we can state precisely the maximum area of various types of town. A pedestrian agglomeration such as we find in China, and such as most of the European towns probably were before the nineteenth century, would be swept entirely by a radius of four miles about the business quarter and industrial centre; and, under these circumstances, where the area of the feeding regions has been very large the massing of human beings has probably reached its extreme limit.* Of course, in the case of a navigable river, for example, the commercial centre might be elongated into a line and the circle of the city modified into an ellipse with a long diameter considerably exceeding eight miles, as, for example, in the case of Hankow.

If, now, horseflesh is brought into the problem, an outer radius of six or eight miles from the centre will define a larger area in which the carriage folk, the hackney users, the omnibus customers, and their domestics and domestic camp followers may live and still be members of the city. Towards that limit London was already probably moving at the accession of Queen Victoria, and it was clearly the absolute limit of urban growth—until locomotive mechanisms capable of more than eight miles an hour could be constructed.

And then there came suddenly the railway and the steamship, the former opening with extraordinary abruptness a series of vast through-routes for trade, the latter enormously increasing the security and economy of the traffic on the old water routes. For a time neither of these inventions was applied to the needs of intra-urban transit at all. For a time they were purely centripetal forces. They worked simply to increase the general volume of trade, to increase, that is, the pressure of population upon the urban centres. As a consequence the social history of the middle and later thirds of the nineteenth century, not simply in England but all over the civilised world, is the history of a gigantic rush of population into the magic radius of—for most people—four miles, to suffer there physical and moral disaster less acute but, finally, far more appalling to the imagination than any famine or pestilence that ever swept the world. Well has Mr. George Gissing named nineteenth-century London in one of his great novels the "Whirlpool," the very figure for the nineteenth-century Great City, attractive, tumultuous, and spinning down to death.

But, indeed, these great cities are no permanent maëlstroms. These new forces, at present still so potently centripetal in their influence, bring with them, nevertheless, the distinct promise of a centrifugal application that may be finally equal to the complete reduction of all our present congestions. The limit of the pre-railway city was the limit of man and horse. But already that limit has been exceeded, and each day brings us nearer to the

*It is worth remarking that in 1801 the density of population in the City of London was half as dense again as that of any district, even of the densest "slum" districts, to-day.

time when it will be thrust outward in every direction with an effect of enormous relief.

So far the only additions to the foot and horse of the old dispensation that have actually come into operation, are the suburban railways, which render possible an average door to office hour's journey of ten or a dozen miles–further only in the case of some specially favoured localities. The star-shaped contour of the modern great city, thrusting out arms along every available railway line, knotted arms of which every knot marks a station, testify sufficiently to the relief of pressure thus afforded. Great Towns before this century presented rounded contours and grew as a puff-ball swells; the modern Great City looks like something that has burst an intolerable envelope and splashed. But, as our previous paper has sought to make clear, these suburban railways are the mere first rough expedient of far more convenient and rapid developments.

We are–as the Census Returns for 1901 quite clearly show–in the early phase of a great development of centrifugal possibilities. And since it has been shown that a city of pedestrians is inexorably limited by a radius of about four miles, and that a horse-using city may grow out to seven or eight, it follows that the available area of a city which can offer a cheap suburban journey of thirty miles an hour is a circle with a radius of thirty miles. And is it too much, therefore, in view of all that has been adduced in this and the previous paper, to expect that the available area for even the common daily toilers of the great city of the year 2000, or earlier, will have a radius very much larger even than that? Now, a circle with a radius of thirty miles gives an area of over 2800 square miles, which is almost a quarter that of Belgium. But thirty miles is only a very moderate estimate of speed, and the reader of the former paper will agree, I think, that the available area for the social equivalent of the favoured season-ticket holders of to-day will have a radius of over one hundred miles, and be almost equal to the area of Ireland.* The radius that will sweep the area available for such as now live in the outer suburbs will include a still vaster area. Indeed, it is not too much to say that the London citizen of the year 2000 A.D. may have a choice of nearly all England and Wales south of Nottingham and east of Exeter as his suburb, and that the vast stretch of country from Washington to Albany will be all of it "available" to the active citizen of New York and Philadelphia before that date.

*Be it noted that the phrase "available area" is used, and various other modifying considerations altogether waived for the present.

This does not for a moment imply that cities of the density of our existing great cities will spread to these limits. Even if we were to suppose the increase of the populations of the great cities to go on at its present rate, this enormous extension of available area would still mean a great possibility of diffusion. But though most great cities are probably still very far from their maxima, though the network of feeding railways has still to spread over Africa and China, and though huge areas are still imperfectly productive for want of a cultivating population, yet it is well to remember that for each great city, quite irrespective of its available spaces, a maximum of population is fixed. Each great city is sustained finally by the trade and production of a certain proportion of the world's surface–by the area it commands commercially. The great city cannot grow, except as a result of some quite morbid and transitory process–to be cured at last by famine and disorder–beyond the limit the commercial capacity of that commanded area prescribes. Long before the population of this city, with its inner circle a third of the area of Belgium, rose towards the old-fashioned city density, this restriction would come in. Even if we allowed for considerable increase in the production of food-stuffs in the future, it still remains inevitable that the increase of each city in the world must come at last upon arrest.

Yet, though one may find reasons for anticipating that this city will in the end overtake and surpass that one and such-like relative prophesying, it is difficult to find any data from which to infer the absolute numerical limits of these various diffused cities. Or perhaps it is more seemly to admit that no such data have occurred to the writer. So far as London, St. Petersburg, and Berlin go, it seems fairly safe to assume that they will go well over twenty millions; and that New York, Philadelphia, and Chicago will probably, and Hankow almost certainly, reach forty millions. Yet even forty millions over thirty-one thousand square miles of territory is, in comparison with four millions over fifty square miles, a highly diffused population.

How far will that possible diffusion accomplish itself? Let us first of all consider the case of those classes that will be free to exercise a choice in the matter, and we shall then be in a better position to consider those more numerous classes whose general circumstances are practically dictated to them. What will be the forces acting upon the prosperous household, the household with a working head and four hundred a year and upwards to live upon, in the days to come? Will the resultant of these forces be, as a rule, centripetal or centrifugal? Will such householders in the greater London of 2000 A.D. still cluster for the most part, as they do to-day, in a group of suburbs as close to London as is compatible with a certain fashionable maximum of garden space and air; or will they leave the ripened gardens and the no longer brilliant villas of Surbiton and Norwood, Tooting and Beckenham, to other and less independent people? First, let us weigh the centrifugal attractions.

The first of these is what is known as the passion for nature, that passion for hillside, wind, and sea that is evident in so many people nowadays, either frankly expressed or disguising itself as a passion for golfing, fishing, hunting, yachting, or cycling; and, secondly, there is the allied charm of cultivation, and especially of gardening, a charm that is partly also the love of dominion, perhaps, and partly a personal love for the beauty of trees and flowers and natural things. Through that we come to a third factor, that craving–strongest, perhaps, in those Low German peoples, who are now ascendant throughout the world–for a little private *imperium* such as a house or cottage "in its own grounds" affords; and from that we pass on to the intense desire so many women feel–and just the women, too, who will mother the future–their almost instinctive demand, indeed, for a household, a separate sacred and distinctive household, built and ordered after their own hearts, such as in its fulness only the country-side permits. Add to these things the healthfulness of the country for young children, and the wholesome isolation that is possible from much that irritates, stimulates prematurely, and corrupts in crowded centres, and the chief positive centrifugal inducements are stated, inducements that no progress of inventions, at any-rate, can ever seriously weaken. What now are the centripetal forces against which these inducements contend?

In the first place, there are a group of forces that will diminish in strength. There is at present the greater convenience of "shopping" within a short radius of the centre of the great city, a very important consideration indeed to many wives and mothers. All the inner and many of the outer suburbs of London obtain an enormous proportion of the ordinary household goods from half a dozen huge furniture, grocery, and drapery firms, each of which has been enabled by the dearness and inefficiency of the parcels distribution of the post-office and railways to elaborate a now very efficient private system of taking orders and delivering goods. Collectively these great businesses have been able to establish a sort of monopoly of suburban trade, to overwhelm the small suburban general tradesman (a fate that was inevitable for him in some way or other), and–which is a positive world-wide misfortune–to overwhelm also many highly specialised shops and dealers of the central district. Suburban people nowadays get their wine and their novels, their clothes and their amusements, their furniture and their food, from some one vast indiscriminate shop or "store" full of respectable mediocre goods, as excellent a thing for housekeeping as it is disastrous to taste and individuality.* But it is doubtful if the delivery organisation of these great stores is any more permanent than the token coinage of the tradespeople of the last century. Just as it was with that interesting development, so now it is with parcels distribution: private enterprise supplies in a partial manner a public need, and with the organisation of a public parcels and goods delivery on cheap and sane lines in the place of our present complex, stupid, confusing, untrustworthy, and fantastically costly chaos of post-office, railways, and carriers, it is quite conceivable that Messrs. Omnium will give place again to specialised shops.

It must always be remembered how timid, tentative, and dear the postal and telephone services of even the most civilised countries still are, and how inexorably the needs of revenue, public profit, and convenience fight in these departments against the tradition of official leisure and dignity. There is no reason now, except that the thing is not yet properly organised, why a telephone call from any point in such a small country as England to any other should cost much more than a postcard. There is no reason now, save railway rivalries and retail ideas–obstacles some able active man is certain to sweep away sooner or later–why the post-office should not deliver parcels anywhere within a radius of a hundred miles in a few hours at a penny or less for a pound and a little over,** put our newspapers in our letter-boxes direct from the printing-office, and, in fact, hand in nearly every constant need of the civilised household, except possibly butcher's meat, coals, greengrocery, and drink. And since there is no reason, but quite removable obstacles, to prevent this development of the post-office, I imagine it will be doing all these things within the next half-century. When it is, this particular centripetal pull, at anyrate, will have altogether ceased to operate.

*Their temporary suppression of the specialist is indeed carried to such an extent that one may see even such things as bronze ornaments and personal jewellery listed in Messrs. Omnium's list, and stored in list designs and pattern; and their assistants will inform you that their brooch, No. 175, is now "very much worn," without either blush or smile.

**The present system of charging parcels by the pound, when goods are sold by the pound, and so getting a miserly profit in the packing, is surely one of the absurdest disregards of the obvious it is possible to imagine.

A second important centripetal consideration at present is the desirability of access to good schools and to the doctor. To leave the great centres is either to abandon one's children, or to buy air for them at the cost of educational disadvantages. But access, be it noted, is another word for transit. It is doubtful if these two needs will so much keep people close to the great city centres as draw them together about secondary centres. New centres they may be–compare Hindhead, for example–in many cases; but also, it may be, in many cases the more healthy and picturesque of the existing small towns will develop a new life. Already, in the case of the London area, such once practically autonomous places as Guildford, Tunbridge Wells, and Godalming have become economically the centres of lax suburbs, and the same fate may very probably overtake, for example, Shrewsbury, Stratford, and Exeter, and remoter and yet remoter townships. Indeed, for all that this particular centripetal force can do, the confluent "residential suburbs" of London, of the great Lancashire-Yorkshire city, and of the Scotch city, may quite conceivably replace the summer lodging-house watering-places of to-day, and extend themselves right round the coast of Great Britain, before the end of the next century, and every open space of mountain and heather be dotted–not too thickly–with clumps of prosperous houses about school, doctor, engineers, book and provision-shops.

A third centripetal force will not be set aside so easily. The direct antagonist it is to that love of nature that drives people out to moor and mountain. One may call it the love of the crowd; and closely allied to it is that love of the theatre which holds so many people in bondage to the Strand. Charles Lamb was the Richard Jefferies of this group of tendencies, and the current disposition to exaggerate the opposition force, especially among English-speaking peoples, should not blind us to the reality of their strength. Moreover, interweaving with these influences that draw people together are other more egotistical and intenser motives, ardent in youth and by no means–to judge by the Folkestone Leas–extinct in age, the love of dress, the love of the crush, the hot passion for the promenade. Here, no doubt, what one may speak of loosely as "racial" characteristics count for much. The common actor and actress of all nationalities, the Neapolitan, the modern Roman, the Parisian, the Hindoo, I am told, and that new and interesting type, the rich and liberated Jew emerging from his Ghetto and free now absolutely to show what stuff he is made of, flame out most gloriously in this direction.

To a certain extent this group of tendencies may lead to the formation of new secondary centres within the "available" area, theatrical and musical centres–centres of extreme Fashion and Selectness, centres of smartness and opulent display–but it is probable that for the large number of people throughout the world who cannot afford to maintain households in duplicate these will be for many years yet strictly centripetal forces, and will keep them within the radius marked by whatever will be the future equivalent in length of, say, the present two-shilling cab ride in London.

And, after all, for all such "shopping" as one cannot do by telephone or postcard, it will still be natural for the shops to be gathered together in some central place. And "shopping" needs refreshment, and may culminate in relaxation. So that Bond Street and Regent Street, the Boulevard des Capucines, the Corso, and Broadway will still be brilliant and crowded for many years for all the diffusion that is here forecast–all the more brilliant and crowded, perhaps, for the lack of a thronging horse-traffic down their central ways. But the very fact that the old nucleus is still to be the best place for all who trade in a concourse of people, for novelty shops and art shops, and theatres and business building, by keeping up the central ground-values will operate against residence there and shift the "masses" outwardly.

And once people have been driven into cab, train, or omnibus, the only reason why they should get out to a residence here rather than there is the necessity of saving time, and such a violent upward gradient of fares as will quite outbalance the downward gradient of ground-values. We have, however, already forecast a swift, varied, and inevitably competitive suburban traffic. And so, though the centre will probably still remain the centre and "Town," it will be essentially a bazaar, a great gallery of shops and places of concourse and rendezvous, a pedestrian place, its pathways reinforced by lifts and moving platforms, and shielded from the weather, and altogether a very spacious, brilliant, and entertaining agglomeration.

Enough now has been said to determine the general nature of the expansion of the great cities in the future, so far as the more prosperous classes are concerned. It will not be a regular diffusion like the diffusion of a gas, but a process of throwing out the "homes," and of segregating various types of people. The omens seem to point pretty unmistakably to a wide and quite unprecedented diversity in the various suburban townships and suburban districts. Of that aspect of the matter a later paper must treat. It is evident that from the outset racial and national characteristics will tell in this diffusion. We are getting near the end of the great Democratic, Wholesale, or Homogeneous phase in the world's history. The sport-loving Englishman, the sociable Frenchman, the vehement American will each diffuse his own great city in his own way.

And now, how will the increase in the facilities of communication we have assumed affect the condition of those whose circumstances are more largely dictated by economic forces? The mere diffusion of a large proportion of the prosperous and relatively free, and the multiplication of various types of road and mechanical traction, means, of course, that in this way alone a perceptible diffusion of the less independent classes will occur. To the subsidiary centres will be drawn doctor and schoolmaster, and various dealers in fresh provisions, baker, grocer, butcher; of if they are already established there they will flourish more and more, and about them the convenient home of the future, with its numerous electrical and mechanical appliances, and the various bicycles, motor-cars, photographic and phonographic apparatus that will be included in its equipment will gather a population of repairers, "accessory" dealers and working engineers, a growing class which from its necessary intelligence and numbers will play a very conspicuous part in the social development of the twentieth century. The much more elaborate post-office and telephone services will also bring intelligent ingredients to these suburban nuclei, these restorations of the old villages and country towns. And the sons of the cottager within the affected area will develop into the skilled vegetable or flower-gardeners, the skilled ostler–with some veterinary science–and so forth, for whom also there will evidently be work and a living. And dotted at every convenient position along the new roads, availing themselves no doubt whenever possible of the picturesque inns that the old coaching days have left us, will be wayside restaurants and tea houses, and motor and cycle stores and repair places. So much diffusion is practically inevitable.

In addition, as we have already intimated, many Londoners in the future may abandon the city office altogether, preferring to do their business in more agreeable surroundings. Such a business as book publishing, for example, has no unbreakable bonds to keep it in the region of high rent and congested streets. The days when the financial fortunes of books depended upon the colloquial support of influential people in a small Society are past; neither publishers nor authors as a class have any relation to Society at all, and actual access to newspaper offices is necessary only to the ranker forms of literary imposture. That personal intercourse between publishers and the miscellaneous race of authors which once justified the central position has, I am told, long since ceased. And the withdrawing publishers may very well take with them the printers and binders, and attract about them their illustrators and designers. . . . So, as a typical instance, one–now urban–trade may detach itself.

Publishing is, however, only one of the many similar trades equally profitable and equally likely to move outward to secondary centres, with the development and cheapening of transit. It is all a question of transit. Limitation of transit contracts the city, facilitation expands and disperses it. All this case for diffusion so far is built up entirely on the hypothesis we attempted to establish in the first paper, that transit of persons and goods alike is to become easier, swifter, and altogether better organised than it is at present.

The telephone will almost certainly prove a very potent auxiliary indeed to the forces making for diffusion. At present that convenience is still needlessly expensive in Great Britain, and a scandalously stupid business conflict between telephone company and post-office delays, complicates, and makes costly and exasperating all trunk communications; but even under these disadvantages the thing is becoming a factor in the life of ordinary villadom. Consider all that lies within its possibilities. Take first the domestic and social side; almost all the labour of ordinary shopping can be avoided–goods nowadays can be ordered and sent either as sold outright, or on approval, to any place within a hundred miles of London, and in one day they can be examined, discussed, and returned–at anyrate, in theory. The mistress of the house has all her local tradesmen, all the great London shops, the circulating library, the theatre box-office, the post-office, and cab-rank, the nurses' institute and the doctor, within reach of her hand. The instrument we may confidently expect to improve, but even now speech is perfectly clear and distinct over several hundred miles of wire. Appointments and invitations can be made; and at a cost varying from a penny to two shillings anyone within two hundred miles of home may speak day or night into the ear of his or her household. Were it not for that unmitigated public nuisance, the practical control of our post-office by non-dismissable Civil servants, appointed so young as to be entirely ignorant of the unofficial world, it would be possible now to send urgent messages at any hour of the day or night to any part of the world; and even our sacred institution of the Civil Service can scarcely prevent this desirable consummation for many years more. The business man may then sit at home in his library and bargain, discuss, promise, hint, threaten, tell such lies as he dare not write, and, in fact, do everything that once demanded a personal encounter. Already for a great number of businesses it is no longer necessary that the office should be in London, and only habit, tradition, and minor considerations keep it there. With the steady cheapening and the steady increase in efficiency of postal and telephonic facilities, and of goods transit, it seems only reasonable to anticipate the need for that expensive office and the irksome daily journey will steadily decline. In other words, what will still be economically the "city," as distinguished from the "agricultural" population, will probably be free to extend, in the case of all the prosperous classes not tied to large establishments in need of personal supervision, far beyond the extreme limits of the daily hour journey.

But the diffusion of the prosperous, independent, and managing classes involves in itself a very considerable diffusion of the purely "working" classes also. Their centres of occupation will be distributed, and their freedom to live at some little distance from their work will be increased. Whether this will mean dotting the country with dull, ugly little streets, slum villages like Buckfastleigh in Devon, for example, or whether it may result in entirely different and novel aspects, is a point for which at present we are not ready. But it bears upon the question that ugliness and squalor upon the main road will appeal to the more prosperous for remedy with far more vigour than when they are stowed compactly in a slum.

Enough has been said to demonstrate that old "town" and "city" will be, in truth, terms as obsolete as "mail coach." For these new areas that will grow out of them we want a term, and the administrative "urban district" presents itself with a convenient air of suggestion. We may for our present purposes call these coming town provinces "urban regions." Practically, by a process of confluence, the whole of Great Britain south of the Highlands seems destined to become such an urban region, laced all together not only by railway and telegraph, but by novel roads such as we forecast in the former chapter, and by a dense network of telephones, parcels-delivery tubes, and the like nervous and arterial connections.

It will certainly be a curious and varied region, far less monotonous than our present English world, still in its thinner regions, at anyrate, wooded, perhaps rather more abundantly wooded, breaking continually into park and garden, and with everywhere a scattering of houses. These will not, as a rule, I should fancy, follow the fashion of the vulgar ready-built villas of the existing suburb, because the freedom people will be able to exercise in the choice of a site will rob the "building estate" promoter of his local advantage; in many cases the houses may very probably be personal homes, built for themselves as much as the Tudor manor-houses were, and even, in some cases, as æsthetically right. Each district, I am inclined to think, will develop its own differences of type and style. As one travels through the urban region, one will traverse open, breezy, "horsey" suburbs, smart white gates and palings everywhere, good turf, a Grand Stand shining pleasantly; gardening districts all set with gables and roses, holly hedges, and emerald lawns; pleasant homes among heathery moorlands and golf-links, and river districts with gaily painted boat-houses peeping from the osiers. Then presently a gathering of houses closer together, and a promenade and a whiff of band and dresses, and then, perhaps, a little island of agriculture, hops, or strawberry gardens, fields of grey-plumed artichokes, white-painted orchard, or brightly neat poultry-farm. Through the varied country the new wide roads will run, here cutting through a crest and there running like some colossal aqueduct across a valley, swarming always with a multitudinous traffic of bright, swift (and not necessarily ugly) mechanisms; and everywhere amidst the fields and trees linking wires will stretch from pole to pole. Ever and again there will appear a cluster of cottages–cottages into which we shall presently look more closely–about some works or workings, works, it may be, with the smoky chimney of to-day replaced by a gaily painted wind-wheel or water-wheel to gather and store the force for the machinery; and ever and again will come a little town, with its cherished ancient church or cathedral, its school buildings and museums, its railway-station, perhaps its fire-station, its inns and restaurants, and with all the wires of the country-side converging to its offices. All that is pleasant and fair of our present country-side may conceivably still be there among the other things. There is no reason why the essential charm of the country should disappear; the new roads will not supersede the present highroads, which will still be necessary for horses and subsidiary traffic; and the lanes and hedges, the field-paths and wild-flowers, will still have their ample justification. A certain lack of solitude there may be perhaps, and–

Will conspicuous advertisements play any part in the landscape? . . .

But I find my pen is running ahead, an imagination prone to

realistic constructions is struggling to paint a picture altogether prematurely. There is very much to be weighed and decided before we can get from our present generalisation to the style of architecture these houses will show, and to the power and nature of the public taste. We have laid down now the broad lines of road, railway, and sea transit in the coming century, and we have got this general prophesy of "urban regions" established, and for the present that much must suffice.

And as for the world beyond our urban regions? The same line of reasoning that leads to the expectation that the city will diffuse itself until it has taken up considerable areas and many of the characteristics, the greenness, the fresh air, of what is now country, leads us to suppose also that the country will take to itself many of the qualities of the city. The old antithesis will indeed cease, the boundary-lines will altogether disappear; it will become, indeed, merely a question of more or less populous. There will be horticulture and agriculture going on within the "urban regions," and "urbanity" without them. Everywhere, indeed, over the land of the globe between the frozen circles, the railway and the new roads will spread, the network of communication-wires and safe and convenient ways. To receive the daily paper a few hours late, to wait a day or so for goods one has ordered, will be the extreme measure of rusticity save in a few remote islands and inaccessible places. The character of the meshes in that wider network of roads that will be the country, as distinguished from the urban district, will vary with the soil, the climate and the tenure of the land—will vary, too, with the racial and national differences. But throughout all that follows, this mere relativity of the new sort of town to the new sort of country over which the new sorts of people we are immediately to consider will be scattered, must be borne in mind.

[At the risk of insistence, I must repeat that, so far, I have been studiously taking no account of the fact that there is such a thing as a boundary-line or a foreigner in the world. It will be far the best thing to continue to do this until we can get out all that will probably happen universally or generally, and in particular the probable changes in social forces, social apparatus and internal political methods. We shall then come to the discussion of language, nationality and international conflicts, equipped with such an array of probabilities and possibilities as will enable us to guess at these special issues with an appearance of far more precision than would be the case if we considered them now.]

III. DEVELOPING SOCIAL ELEMENTS

THE mere differences in thickness of population and facility of movement that have been discussed thus far, will involve consequences remarkable enough, upon the *facies* of the social body; but there are certain still broader features of the social order of the coming time, less intimately related to transit, that it will be convenient to discuss at this stage. They are essentially outcomes of the enormous development of mechanism which has been the cardinal feature of the nineteenth century; for this development, by altering the method and proportions of almost all human undertakings,* has altered absolutely the grouping and character of the groups of human beings engaged upon them.

Throughout the world for forty centuries the more highly developed societies have always presented under a considerable variety of superficial differences certain features in common. Always at the base of the edifice, supporting all, subordinate to all, and the most necessary of all, there has been the working cultivator, peasant, serf, or slave. Save for a little water-power, a little use of windmills, the traction of a horse or mule, this class has been the source of all the work upon which the community depends. And, moreover, whatever labour town developments have demanded has been supplied by the muscle of its fecund ranks. It has been, in fact–and to some extent still is–the multitudinous living machinery of the old social order; it carried, cropped, tilled, built, and made. And, directing and sometimes owning this human machinery, there has always been a superior

*Even the characteristic conditions of writing books, that least mechanical of pursuits, have been profoundly affected by the typewriter.

class, bound usually by a point of honour not to toil, often warlike, often equestrian, and sometimes cultivated. In England this is the gentility, in most European countries it is organised as a nobility; it is represented in the history of India by the "twice born" castes, and in China–the most philosophically conceived and the most stably organised social system the old order ever developed–it finds its equivalent in the members of a variously buttoned mandarinate, who ride, not on horses, but on a once adequate and still respectable erudition. These two primary classes may and do become in many cases complicated by subdivisions; the peasant class may split into farmers and labourers, the gentlemen admit a series of grades and orders, kings, dukes, earls, and the like, but the broad distinction remains intact, as though it was a distinction residing in the nature of things.*

From the very dawn of history until the first beginnings of mechanism in the eighteenth century, this simple scheme of orders was the universal organisation of all but savage humanity, and the chief substance of history until these later years has been in essence the perpetual endeavour of specific social systems of this type to attain in every region the locally suitable permanent form, in face of those two inveterate enemies of human stability, innovation, and that secular increase in population that security permits. The imperfection of the means of communication rendered political unions of a greater area than that swept by a hundred-mile radius highly unstable. It was a world of small states. Lax empires came and went, at the utmost they were the linking of practically autonomous states under a common *Pax*. Wars were usually wars between kingdoms, conflicts of this local experiment in social organisation with that. Through all the historical period these two well-defined classes of gentle and simple acted and reacted upon each other, every individual in each class driven by that same will to live and do, that imperative of self-establishment and aggression that is the spirit of this world. Until the coming of gunpowder, the man on horseback–commonly with some sort of armour–was invincible in battle in the open. Wherever the land lay wide and unbroken, and the great lines

*To these two primary classes the more complicated societies have added others. There is the priest, almost always in the social order of the pre-railway period, an integral part, a functional organ of the social body, and there are the lawyer and the physician. And in the towns–constituting, indeed, the towns–there appear, as an outgrowth of the toiling class, a little emancipated from the gentleman's direct control, the craftsman, the merchant, and the trading sailor, essentially accessory classes, producers of, and dealers in, the accessories of life, and mitigating and clouding only very slightly that broad duality.

of trade did not fall, there the horseman was master–or the clerkly man behind the horseman. Such a land was aristocratic and tended to form castes. The craftsman sheltered under a patron, and in guilds in a walled town, and the labourer was a serf. He was ruled over by knight or by his creditor–in the end it matters little how the gentleman began. But where the land became difficult by reason of mountain or forest, or where water greatly intersected it, the pikeman or close-fighting swordsman or the bowman could hold his own, and a democratic flavour, a touch of repudiation, was in the air. In such countries as Italy, Greece, the Alps, the Netherlands, and Great Britain, the two forces of the old order, the aristocrat and the common man, were in a state of unstable equilibrium, through the whole period of history. A slight change* in the details of the conflict for existence could tilt the balance. A weapon a little better adapted to one class than the other, or a slight widening of the educational gap, worked out into historically imposing results, to dynastic changes, class revolutions and the passing of empires.

Throughout it was essentially one phase of human organisation. When one comes to examine the final result, it is astonishing to remark the small amount of essential change, of positively final and irreparable alternation, in the conditions of the common life. Consider, for example, how entirely in sympathy was the close of the eighteenth century with the epoch of Horace, and how closely equivalent were the various social aspects of the two periods. The literature of Rome was living reading in a sense that has suddenly passed away, it fitted all occasions, it conflicted with no essential facts in life. It was a commonplace of the thought of that time that all things recurred, all things circled back to their former seasons; there was nothing new under the sun. But now almost suddenly the circling has ceased, and we find ourselves breaking away. Correlated with the sudden development of mechanical forces that first began to be socially perceptible in the middle eighteenth century, has been the appearance of great masses of population, having quite novel functions and relations in the social body, and together with this appearance such a suppression, curtailment, and modification of the older classes, as to point to an entire disintegration of that system. The *facies* of the social fabric has changed, and–as I hope to make clear–is still changing in a direction from which, without a total destruction and rebirth of that fabric, there can never be any return.

*Slight, that is, in comparison with nineteenth-century changes.

The most striking of the new classes to emerge is certainly the shareholding class, the owners of a sort of property new in the world's history.

Before the eighteenth century the only property of serious importance consisted of land and buildings. These were "real" estate. Beyond these things were live-stock, serfs, and the furnishings of real estate, the surface aspect of real estate, as to speak, personal property, ships, weapons, and the Semitic invention of money. All such property had to be actually "held" and administered by the owner, he was immediately in connection with it and responsible for it. He could leave it only precariously to a steward and manager, and to convey the revenue of it to him at a distance was a difficult and costly proceeding. To prevent a constant social disturbance by lapsing and dividing property, and in the absence of any organised agency to receive lapsed property, inheritance and preferably primogeniture were of such manifest advantage that the old social organisation always tended in the direction of these institutions. Such usury as was practiced relied entirely on the land and the anticipated agricultural produce of the land.

But the usury and the sleeping partnerships of the Joint Stock Company system which took shape in the eighteenth and the earlier half of the nineteenth century opened quite unprecedented uses for money, and created a practically new sort of property and a new proprietor class. The peculiar novelty of this property is easily defined. Given a sufficient sentiment of public honesty, share property is property that can be owned at any distance and that yields its revenue without thought or care on the part of its proprietor; it is, indeed, absolutely irresponsible property, a thing that no old world property ever was. But, in spite of its widely different nature, the laws of inheritance that the social necessities of the old order of things established have been applied to this new species of possession without remark. It is indestructible, imperishable wealth, subject only to the mutations of value that economic changes bring about. Related in its character of absolute irresponsibility to this shareholding class is a kindred class that has grown with the growth of the great towns, the people who live upon ground-rents. There is every indication that this element of irresponsible, independent, and wealthy people in the social body, people who feel the urgency of no exertion, the pressure of no specific positive duties, is still on the increase, and may still for a long time increasingly preponderate. It overshadows the responsible owner of the real property or of real businesses altogether. And most of the old aristocrats, the old knightly and landholding people, have, so to speak, converted themselves into members of this new class.

It is a class with scarcely any specific characteristics beyond its defining one, of the possession of property and all the potentialities property entails, with a total lack of function with regard to that property. It is not even collected into a distinct mass. It graduates insensibly into every other class, it permeates society as threads and veins of gold permeate quartz. It includes the millionaire snob, the political-minded plutocrat, the wealthy sensualist, open-handed religious fanatics, the "Charitable," the smart, the magnificently dull, the great army of timid creatures who tremble through life on a safe bare sufficiency,* travellers, hunters, minor poets, sporting enthusiasts, many of the officers in the British Army, and all sorts and conditions of amateurs. In a sense it includes several modern royalties, for the crown in several modern constitutional states is a *corporation sole,* and the monarch the unique, unlimited, and so far as necessity goes, quite functionless shareholder. He may be a heavy-eyed sensualist, a small-minded leader of fashion, a rival to his servants in the gay science of etiquette, a frequenter of race-courses and music-halls, a literary or scientific quack, a devotee, an amateur anything–the point is that his income and sustenance have no relation whatever to his activities. If he fancies it, or is urged to it by those who have influence over him, he may even "be a king!" But that is not compulsory, not essential, and there are practically no conditional restrictions whatever laid upon him.

Those who belong to this shareholding class only partially, who partially depend upon dividends and partially upon activities, occur in every rank and order of the whole social body. The waiter one tips probably has a hundred or so in some remote company, the will of the eminent labour reformer reveals an admirably distributed series of investments, the bishop sells tea and digs coal, or at anyrate gets a profit from some unknown persons tea-selling or coal-digging, to eke out the direct recompense of his own modest corn-treading. Indeed, above the labouring class, the number of individuals in the social body whose gross income is entirely the result of their social activities is very small. Previously in the world's history, saving a few quite exceptional aspects, the possession and retention of property was conditional upon activities of some sort, honest or dishonest, work, force, or fraud. But the shareholding ingredient of

*It included, one remembers, Schopenhauer, but, as he remarked upon occasion, not Hegel.

our new society, so far as its shareholding goes, has no need of strength or wisdom; the countless untraceable Owner of the modern world presents in a multitudinous form the image of a Merovingian king. The shareholder owns the world *de jure,* by the common recognition of the rights of property; and the incumbency of knowledge, management, and toil fall entirely to others. He toils not, neither does he spin; he is mechanically released from the penalty of the Fall, he reaps in a still sinful world all the practical benefits of a millennium–without any of its moral limitations.

It will be well to glance at certain considerations which point to the by no means self-evident proposition, that this factor of irresponsible property is certain to be present in the social body a hundred years ahead. It has, no doubt, occurred to the reader that all the conditions of the shareholder's being unfit him for co-operative action in defence of the interests of his class. Since shareholders do nothing in common, except receive and hope for dividends, since they may be of any class, any culture, any disposition, or any level of capacity, since there is nothing to make them read the same papers, gather in the same places, or feel any sort of sympathy with each other beyond the universal sympathy of man for man, they will, one may anticipate, be incapable of any concerted action to defend the income they draw from society against any resolute attack. Such crude and obvious denials of the essential principles of their existence as the various Socialistic bodies have proclaimed have, no doubt, encountered a vast, unorganised, negative opposition from them, but the subtle and varied attack of natural forces they have neither the collective intelligence to recognise, nor the natural organisation to resist. The shareholding body is altogether too chaotic and diffused for positive defence. And the question of the prolonged existence of this comparatively new social phenomenon, either in its present or some modified form, turns, therefore, entirely on the quasi-natural laws of the social body. If they favour it, it will survive; when they do not, it will vanish as the mists of the morning before the sun.

Neglecting a few exceptional older corporations which, indeed, in their essence are not usurious, but of unlimited liability, the shareholding body appeared first, in its present character, in the seventeenth century, and came to its full development in the mid-nineteenth. Was its appearance then due only to the attainment of a certain necessary degree of public credit, or was it correlated with any other force? It seems in accordance with facts to relate it to another force, the development of mechanism, so far as certain representative aspects go. Hitherto the only borrower had been the farmer, then the exploring trader had found a world too wide for purely individual effort, and then suddenly the craftsmen of all sorts and the carriers discovered the need of the new, great, wholesale, initially expensive appliances that invention was offering them. It was the development of mechanism that created the great bulk of modern shareholding, it took its present shape distinctively only with the appearance of the railways. The hitherto necessary but subordinate craftsman and merchant classes were to have new weapons, new powers, they were to develop to a new importance, to a preponderance even in the social body. But before they could attain these weapons, before this new and novel wealth could be set up, it had to pay its footing in an apportioned world, it had to buy its right to disturb the established social order. The dividend of the shareholder was the tribute the new enterprise had to pay the old wealth. The share was the manumission money of machinery. And essentially the shareholder represents and will continue to represent the responsible managing owner of a former state of affairs in process of supersession.

If the great material developments of the nineteenth century had been final, if they had, indeed, constituted merely a revolution and not an absolute release from the fixed conditions about which human affairs circled, we might even now be settling accounts with our Merovingians as the socialists desire. But these developments were not final, and one sees no hint as yet of any coming finality. Invention runs free and our state is under its dominion. The novel is continually struggling to establish itself at the relative or absolute expense of the old. The statesman's conception of social organisation is no longer stability but growth. And so long as material progress continues, this tribute must continue to be paid; so long as the stream of development flows, this necessary back eddy will endure. Even if we "municipalise" all sorts of undertakings we shall not alter the essential facts, we shall only substitute for the shareholder the corporation stockholder. The figure of an eddy is particularly appropriate. Enterprises will come and go, the relative values of kinds of wealth will alter, old appliances, old companies, will serve their time and fall in value, individuals will waste their substance, individual families and groups will die out, certain portions of the share property of the world may be gathered, by elaborate manipulation, into a more or less limited number of hands, conceivably even families and groups will be taxed out by graduated legacy-duties and specially apportioned income-taxes, but, for all such possible changes and modifications, the shareholding element will still endure, so long as our present progressive and experimental state of society obtains. And the very diversity, laxity, and weakness of the general shareholding element, which will work to prevent its organising itself in the interests of its property, or of evolving any distinctive traditions or positive characters, will obviously prevent its obstructing the continual appearance of new enterprises, of new shareholders to replace the loss of its older constituents. . . .

At the opposite pole of the social scale to that about which shareholding is most apparent, is a second necessary and quite inevitable consequence of the sudden transition that has occurred from a very nearly static social organisation to a violently progressive one. This second consequence of progress is the appearance of a great number of people without either property or any evident function in the social organism. This new ingredient is most apparent in the towns, it is frequently spoken of as the Urban Poor, but its characteristic traits are to be found also in the rural districts. For the most part its individuals are either criminal, immoral, parasitic in more or less irregular ways upon the more or less irregular ways upon the more successful classes, or labouring, at something less than a regular bare subsistence wage, in a finally hopeless competition against machinery that is as yet not so cheap as their toil. It is, to borrow a popular phrase, the "submerged" portion of the social body, a leaderless, aimless multitude, a multitude of people drifting down towards the abyss. Essentially it consists of people who have failed to "catch on" to the altered necessities the development of mechanism has brought about, they are people thrown out of employment by machinery, thrown out of employment by the escape of industries along some newly opened line of communication to some remote part of the world, or born under circumstances that give them no opportunity of entering the world of active work. Into this welter of machine-superseded toil there topples the non-adaptable residue of every changing trade; its members marry and are given in marriage, and it is recruited by the spendthrifts, weaklings, and failures of every superior class.

Since this class was not apparent in masses in the relatively static, relatively less eliminatory, society of former times, its appearance has given rise to a belief that the least desirable section of the community has become unprecedentedly prolific, that there is now going on a "Rapid Multiplication of the Unfit."

But sooner or later, as every East End doctor knows, the ways of the social abyss lead to death, the premature death of the individual, or death through the death or infertility of the individual's stunted offspring, or death through that extinction which moral perversion involves. It is a recruited class, not a breeding multitude. Whatever expedients may be resorted to, to mitigate or conceal the essential nature of this social element, it remains in its essence wherever social progress is being made, the contingent of death. Humanity has set out in the direction of a more complex and exacting organisation, and until, by a foresight to me at least inconceivable, it can prevent the birth of just all the inadaptable, useless, or merely unnecessary creatures in each generation, there must needs continue to be, in greater or less amount, this individually futile struggle beneath the feet of the race; somewhere and in some form there must still persist those essentials that now take shape as the slum, the prison, and the asylum. All over the world, as the railway network has spread, in Chicago and New York as vividly as in London or Paris, the commencement of the new movement has been marked at once by the appearance of this bulky irremovable excretion, the appearance of these gall-stones of vicious, helpless, and pauper masses. There seems every reason to suppose that this phenomenon of unemployed citizens, who are, in fact, unemployable, will remain present as a class, perishing individually and individually renewed, so long as civilisation remains progressive and experimental upon its present lines. Their drowning existences may be utilised, the crude hardship of their lot may be concealed or mitigated,* they may react upon the social fabric that is attempting to eliminate them, in very astounding ways, but their presence and their individual doom, it seems to me, will be unavoidable–at anyrate, for many generations of men. They are an integral part of this physiological process of mechanical progress, as inevitable in the social body

*A very important factor in this mitigation, a factor over which the humanely minded cannot too greatly rejoice, will be the philanthropic amusements of the irresponsible wealthy. There is a growing class of energetic people–organisers, secretaries, preachers–who cater to the philanthropic instinct, and who are, for all practical purposes, employing a large and increasing section of suitable helpless people, in supplying to their customers, by means of religious acquiescence and light moral reforms, that sense of well-doing which is one of the least objectionable of the functionless pleasures of life. The attempts to reinstate these failures by means of subsidised industries will, in the end, of course, merely serve to throw out of employment other just subsisting strugglers; it will probably make little or no difference in the net result of the process.

as are waste matters and disintegrating cells in the body of an active and healthy man.

The appearance of these two strange functionless elements, although the most striking symptom of the new phase of progressive mechanical civilisation now beginning, is by no means the most essential change in progress. These appearances involve also certain disappearances. I have already indicated pretty clearly that the vast irregular development of irresponsible wealthy people is swallowing up and assimilating more and more the old class of administrative land-owning gentlemen in all their grades and degrees. The old upper class, as a functional member of the State, is being effaced. And I have also suggested that the old lower class, the broad necessary base of the social pyramid, the uneducated inadaptable peasants and labourers, is, with the development of toil-saving machinery, dwindling and crumbling down bit by bit towards the abyss. But side by side with these two processes is a third process of still profounder significance, and that is the reconstruction and the vast proliferation of what constituted the middle class of the old order. It is now, indeed, no longer a middle class at all. Rather all the definite classes in the old scheme of functional precedence have melted and mingled,* and in the molten mass there has appeared a vast intricate confusion of different sorts of people, some sailing about upon floating masses of irresponsible property, some buoyed by smaller fragments, some clinging desperately enough to insignificant atoms, a great and varied multitude swimming successfully without aid, or with an amount of aid that is negligible in relation to their own efforts, and an equally varied multitude of less capable ones clinging to the swimmers, clinging to the floating rich, or clutching empty-handed and thrust and sinking down. This is the typical aspect of the modern community. It will serve as a general description of either the United States or any western European State, and the day is not far distant when the extension of means of communication, and of the shareholding method of conducting affairs, will make it applicable to the whole world. Save, possibly, in a few islands and inaccessible places and regardless of colour or creed, this process of deliquescence seems destined to spread. In a great diversity of tongues, in the phases of a number of conflicting moral and theological traditions, in the varying tones of contrasting racial temperaments, the grandchildren of black and white, and red and brown, will

*I reserve any consideration of the special case of the "priest."

be seeking more or less consciously to express themselves in relation to these new and unusual social conditions. But the change itself is no longer amenable to their interpretations, the world-wide spreading of swift communication, the obliteration of town and country, the deliquescence of the local social order, have an air of being processes as uncontrollable by such collective intelligence as men can at present command, and as indifferent to his local peculiarities and prejudices as the movements of winds and tides. . . .

It will be obvious that the interest of this speculation, at anyrate, centres upon this great intermediate mass of people who are neither passively wealthy, the sleeping partners of change, nor helplessly thrust out of the process. Indeed, from our point of view–an inquiry into coming things–these non-effective masses would have but the slightest interest were it not for their enormous possibilities of reaction upon the really living portion of the social organism. This really living portion seems at first sight to be as deliquescent in its nature, to be drifting down to as chaotic a structure as either the nonfunctional owners that float above it or the unemployed who sink below. What were once the definite subdivisions of the middle class modify and lose their boundaries. The retail tradesman of the towns, for example–once a fairly homogeneous class throughout Europe–expands here into vast store companies, and dwindles there to be an agent or collector, seeks employment or topples outright into the abyss. But under a certain scrutiny one can detect here what we do not detect in our other two elements, and that is that, going on side by side with the processes of dissolution and frequently masked by these, there are other processes by which men, often of the most diverse parentage and antecedent traditions, are being segregated into a multitude of specific new groups which may presently develop very distinctive characters and ideals.

There are, for example, the unorganised myriads that one can cover by the phrase "mechanics and engineers," if one uses it in its widest possible sense. At present it would be almost impossible to describe such a thing as a typical engineer, to predicate any universally applicable characteristic of the engineer and mechanic. The black-faced, oily man one figures emerging from the engine-room serves well enough, until one recalls the sanitary engineer with his additions of crockery and plumbing, the electrical engineer with his little tests and wires, the mining engineer, the railway maker, the motor builder, and the irrigation expert. Even if we take some specific branch of all this huge mass of new employment the coming of mechanism has brought with it, we still find an undigested miscellany. Consider the rude levy that is engaged in supplying and repairing the world's new need of bicycles! Wheelwrights, watchmakers, blacksmiths, music-dealers, drapers, sewing-machine repairers, smart errand-boys, ironmongers, individuals from all the older aspects of engineering, have been caught up by the new development, are all now, with a more or less inadequate knowledge and training, working in the new service. But is it likely that this will remain a rude levy? From all these varied people the world requires certain things, and a failure to obtain them involves, sooner or later, in this competitive creation, an individual replacement and a push towards the abyss. The very lowest of them must understand the machine they contribute to make and repair, and not only is it a fairly complex machine in itself, but it is found in several types and patterns, and so far it has altered, and promises still to alter, steadily, by improvements in this part and that. No limited stock-in-trade of knowledge, such as suffices for a joiner or an ostler, will serve. They must keep on mastering new points, new aspects, they must be intelligent and adaptable, they must get a grasp of that permanent something that lies behind the changing immediate practice. In other words, they will have to be educated rather than trained after the fashion of the old craftsman. Just now this body of irregulars is threatened by the coming of the motors. The motors promise new difficulties, new rewards, and new competition. It is an ill look-out for the cycle mechanic who is not prepared to tackle the new problems that will arise. For all this next century this particular body of mechanics will be picking up new recruits and eliminating the incompetent and the rule-of-thumb sage. Can it fail, as the years pass, to develop certain general characters, to become so far homogeneous as to be generally conscious of the need of a scientific education, at anyrate in mechanical and chemical matters, and to possess, down to its very lowest ranks and orders, a common fund of intellectual training?

But the makers and repairers of cycles, and that larger multitude that will presently be concerned with motors, are, after all, only a small and specialised section of the general body of mechanics and engineers. Every year, with the advance of invention, new branches of activity, that change in their nature and methods all too rapidly for the establishment of rote and routine workers of the old type, call together fresh levies of amateurish workers and learners who must surely presently develop into, or give place to, bodies of qualified and capable men. And the point

I would particularly insist upon here is, that throughout all its ranks and ramifications, from the organising heads of great undertakings down to the assistant in the local repair-shop, this new, great, and expanding body of mechanics and engineers will tend to become an educated and adaptable class in a sense that the craftsmen of former times were not educated and adaptable. Just how high the scientific and practical education may rise in the central levels of this body is a matter for subsequent speculation, just how much initiative will be found in the lowest ranks depends upon many very complex considerations. But that here we have at least the possibility, the primary creative conditions of a new, numerous, intelligent, educated, and capable social element is, I think, a proposition with which the reader will agree.

What are the chief obstacles in the way of the emergence, from out the present chaos, of this social element equipped, organised, educated, conscious of itself and of distinctive aims, in the next hundred years? In the first place there is the spirit of trade unionism, the conservative contagion of the old craftsmanship. Trade Unions arose under the tradition of the old order, when in every business, employer and employed stood in marked antagonism, stood as a special instance of the universal relationship of gentle or intelligent, who supplied no labour, and simple, who supplied nothing else. The interest of the employer was to get as much labour as possible out of his hirelings; the complementary object in life of the hireling, whose sole function was drudgery, who had no other prospect until death, was to give as little to his employer as possible. In order to keep the necessary labourer submissive, it was a matter of public policy to keep him uneducated and as near the condition of a beast of burden as possible, and in order to keep his life tolerable against that natural increase which all the moral institutions of his state promoted, the labourer–stimulated if his efforts slackened by the touch of absolute misery–was forced to devise elaborate rules for restricting the hours of toil, making its performance needlessly complex, and shirking with extreme ingenuity and conscientiousness. In the older trades, of which the building-trade is foremost, these two traditions, reinforced by unimaginative building regulations, have practically arrested any advance whatever.* There can be no doubt that this influence has spread

*I find it incredible that there will not be a sweeping revolution in the methods of building during the next century. The erection of a house-wall, come to think of it, is an astonishingly tedious and complex business; the final result exceedingly unsatisfactory. It has been my lot recently to follow in detail the

into what are practically new branches of work. Even where new conveniences have called for new types of workmen and have opened the way for the elevation of a group of labourers to the

process of building a private dwelling-house, and the solemn succession of deliberate, respectable, perfectly satisfied men, who have contributed each so many days of his life to this accumulation of weak compromises, has enormously intensified my constitutional amazement at my fellow-creatures. The chief ingredient in this particular house-wall is the common brick, burnt earth, and but one step from the handfuls of clay of the ancestral mud hut, small in size and permeable to damp. Slowly, day by day, the walls grew tediously up, to a melody of tinkling trowels. These bricks are joined by mortar, which is mixed in small quantities, and must vary very greatly in its quality and properties throughout the house. In order to prevent the obvious evils of a wall of porous and irregular baked clay and lime mud, a damp course of tarred felt, which cannot possibly last more than a few years, was inserted about a foot from the ground. Then the wall, being quite insufficient to stand the heavy drift of weather to which it is exposed, was dabbled over with two coatings of plaster on the outside, the outermost being given a primitive picturesqueness by means of a sham surface of rough-cast pebbles and whitewash, while within, to conceal the rough discomfort of the surface, successive coatings of plaster, and finally, paper, were added, with a wood-skirting at the foot thrice painted. Everything in this was hand-work, the laying of the bricks, the dabbing of the plaster, the smoothing of the paper; it is a house built of hands–and some I saw were bleeding hands–just as in the days of the pyramids, when the only engines were living men. The whole confection is now undergoing incalculable chemical reactions between its several parts. Lime, mortar, and microscopical organisms are producing undesigned chromatic effects in the paper and plaster; the plaster, having methods of expansion and contraction of its own, crinkles and cracks; the skirting, having absorbed moisture and now drying again, opens its joints; the rough-cast coquettes with the frost and opens chinks and crannies for the humbler creation. I fail to see the necessity of (and, accordingly, I resent bitterly) all these coral-reef methods. Better walls than this, and better and less life-wasting ways of making them, are surely possible. In the wall in question, concrete would have been cheaper and better than bricks if only "the men" had understood it. But I can dream at last of much more revolutionary affairs, of a thing running to and fro along a temporary rail, that will squeeze out wall as one squeezes paint from a tube, and form its surface with a pat or two as it sets. Moreover, I do not see at all why the walls of small dwelling-houses should be so solid as they are. There still hangs about us the monumental traditions of the pyramids. It ought to be possible to build sound, portable, and habitable houses of felted wire-netting and weatherproofed paper upon a light framework. This sort of thing is, no doubt, abominably ugly at present, but that is because architects and designers, being for the most part inordinately cultured and quite uneducated, are unable to cope with its fundamentally novel problems. A few energetic men might at any time set out to alter all this. And with the inevitable revolutions that must come about in domestic fittings, and which I hope to discuss more fully in the next paper, it is open to question whether many ground landlords may not find they have work for the house-breakers rather than wealth unlimited falling into their hands when the building leases their solicitors so ingeniously draw up do at last expire.

higher level of educated men,* the old traditions have to a very large extent prevailed. The average sanitary plumber of to-day in England insists upon his position as a mere labourer as though it were some precious thing, he guards himself from improvement as a virtuous woman guards her honour, he works for specifically limited hours and by the hour with specific limitations in the practice of his trade, on the fairly sound assumption that but for that restriction any fool might do plumbing as well as he; whatever he learns he learns from some other plumber during his apprenticeship years–after which he devotes himself to doing the minimum of work in the maximum of time until his brief excursion into this mysterious universe is over. So far from invention spurring him onward, every improvement in sanitary work in England, at least, is limited by the problem whether "the men" will understand it. A person ingenious enough to exceed this sacred limit might as well hang himself as trouble about the improvement of plumbing.

If England stood alone, I do not see why each of the new mechanical and engineering industries, so soon as it develops sufficiently to have gathered together a body of workers capable of supporting a Trade Union secretary, should not begin to stagnate in the same manner. Only England does not stand alone, and the building-trade is so far not typical, inasmuch as it possesses a national monopoly that the most elaborate system of protection cannot secure any other group of trades. One must have one's house built where one has to live, the importation of workmen in small bodies is difficult and dear, and if one cannot have the house one wishes, one must needs have the least offensive substitute; but bicycle and motor, ironwork and furniture, engines, rails, and ships one can import. The community, therefore, that does least to educate its mechanics and engineers out of the base and servile tradition of the old idea of industry will in the coming years of progress simply get a disproportionate share of the rejected element, the trade will go elsewhere, and the community will be left in possession of an exceptionally large contingent for the abyss.

At present, however, I am dealing not with the specific community, but with the generalised civilised community of A.D. 2000–we disregard the fate of states and empires for a time–and, for that emergent community, wherever it may be, it seems

*The new aspects of building, for example, that have been brought about by the entrance of water and gas into the house, and the application of water to sanitation.

reasonable to anticipate, replacing and enormously larger and more important than the classes of common workmen and mechanics of to-day, a large, fairly homogeneous body–big men and little men, indeed, but with no dividing lines–of more or less expert mechanics and engineers, with a certain common minimum of education and intelligence, and probably a common-class consciousness–a new body, a new force, in the world's history.

For this body to exist implies the existence of much more than the primary and initiating nucleus of engineers and skilled mechanics. If it is an educated class, its existence implies a class of educators, and just as far as it does get educated the schoolmasters will be skilled and educated men. The shabby-genteel middle-class schoolmaster of the England of to-day, in–or a little way out of–orders, with his smattering of Greek, his Latin that leads nowhere, his fatuous mathematics, his gross ignorance of pedagogics, and his incomparable snobbishness, certainly does not represent the schoolmaster of this coming class. Moreover, the new element will necessarily embody its collective, necessarily distinctive, and unprecedented thoughts in a literature of its own, its development means the development of a new sort of writer and of new elements in the press. And since, if it does emerge, a revolution in the common schools of the community will be a necessary part of the process, then its emergence will involve a revolutionary change in the condition of classes that might otherwise remain as they are now–the older craftsman, for example.

The process of attraction will not end even there; the development of more and more scientific engineering and of really adaptable operatives will render possible agricultural contrivances that are now only dreams, and the diffusion of this new class over the country-side–assuming the reasoning in my second chapter to be sound–will bring the lever of the improved schools under the agriculturist. The practically autonomous farm of the old epoch will probably be replaced by a great variety of types of cultivation, each with its labour-saving equipment. In this, as in most things, the future spells variation. The practical abolition of impossible distances over the world will tend to make every district specialise in the production for which it is best fitted, and to develop that production with an elaborate precision and economy. The chief opposing force to this tendency will be found in those countries where the tenure of the land is in small holdings. A population of small agriculturists that has really got itself well established is probably as hopelessly immovable a thing as the forces of progressive change will have to encounter. The Arcadian healthiness and simplicity of the small holder, and the usefulness of little hands about him, naturally results in his keeping the population on his plot up to the limit of bare subsistence. He avoids over-education, and his beasts live with him and his children in a natural kindly manner. He will have no idlers, and even grandmamma goes weeding. His net produce is less than the production of the larger methods, but his gross is greater, and usually it is mortgaged more or less. Along the selvage of many of the new roads we have foretold, his hens will peck and his children beg, far into the coming decades. This simple, virtuous, open-air life is to be found ripening in the north of France and Belgium, it culminated in Ireland in the famine years, it has held its own in China–with a use of female infanticide–for immemorable ages, and a number of excellent persons are endeavouring to establish it in England at the present time. At the Cape of Good Hope, under British rule, Kaffirs are being settled upon little inalienable holdings that must inevitably develop in the same direction, and over the Southern States the nigger squats and multiplies. It is fairly certain that these stagnant ponds of population, which will grow until public intelligence rises to the pitch of draining them, will on a greater scale parallel in the twentieth century the soon-to-be-dispersed urban slums of the nineteenth. But I do not see how they can obstruct, more than locally, the reorganisation of agriculture and horticulture upon the ampler and more economical lines mechanism permits, or prevent the development of a type of agriculturist as adaptable, alert, intelligent, unprejudiced, and modest as the coming engineer.

Another great section of the community, the military element, will also fall within the attraction of this possible synthesis, and will inevitably undergo profound modification. Of the probable development of warfare a later chapter shall treat, and here it will suffice to point out that at present science stands proffering the soldier vague, vast possibilities of mechanism, and, so far, he has accepted practically nothing but rifles which he cannot sight and guns that he does not learn to move about. It is quite possible the sailor would be in the like case, but for the exceptional conditions that begot ironclads in the American Civil War. Science offers the soldier transport that he does not use, maps he does not use, entrenching devices, road-making devices, balloons and flying scouts, portable foods, security from disease, a thousand ways of organising the horrible uncertainties of war. But the soldier of to-day–I do not mean the British soldier only–still insists on regarding these revolutionary appliances as mere accessories, and untrustworthy ones at that, to the time-honoured practice of his art. He guards his technical innocence like a plumber.

Every European army is organised on the lines of the once fundamental distinction of the horse and foot epoch, in deference to the contrast of gentle and simple. There is the officer, with all the traditions of old nobility, and the men still, by a hundred implications, mere sources of mechanical force, and fundamentally base. The British Army, for example, still cherishes the tradition that its privates are absolutely illiterate, and such small instruction as is given them in the art of war is imparted by bawling and enforced by abuse upon public drill-grounds. Almost all discussion of military matters still turns upon the now quite stupid assumption that there are two primary military arms and no more, horse and foot. "Cyclists are infantry," the War office manual of 1900 gallantly declares in the face of this changing universe. After fifty years of railways, there still does not exist, in a world which is said to be over devoted to military affairs, a skilled and organised body of men, specially prepared to seize, repair, reconstruct, work, and fight such an important element in the new social machinery as a railway-system. Such a business, in the next European war, will be hastily entrusted to some haphazard incapables drafted from one or other of the two prehistoric arms. . . . I do not see how this condition of affairs can be anything but transitory. There may be several wars between European powers, prepared and organised to accept the old conventions, bloody, vast, distressful encounters that may still leave the art of war essentially unmodified, but sooner or later–it may be in the improvised struggle that follows the collapse of some one of these huge, witless, fighting forces–the new sort of soldier will emerge, a sober, considerate, engineering man–no more of a gentleman than the man subordinated to him or any other self-respecting person. . . .

Certain interesting side-questions I may glance at here, only for the present, at least, to set them aside unanswered, the reaction, for example, of this probable development of a great mass of educated and intelligent efficients upon the status and quality of the medical profession, and the influence of its novel needs in either modifying the existing legal body or calling into being a parallel body of more expert and versatile guides and assistants in business operations. But from the mention of this latter section one comes to another possible centre of aggregation in the social welter. Opposed in many of their most essential conditions to the capable men who are of primary importance in the social body, is the great and growing variety of non-productive but active men who are engaged in more or less necessary operations of organisation, promotion, advertisement, and trade. There are the business managers, public and private, the political organisers, brokers, commission-agents, the varying grades of financier down to the mere greedy camp-followers of finance, the gamblers pure and simple, and the great body of their dependent clerks, typewriters, and assistants. All this multitude will have this much in common, that it will be dealing, not with the primary inexorable logic of natural laws, but with the shifting, uncertain prejudices and emotions of the general mass of people. It will be wary and cunning rather than deliberate and intelligent, smart rather than prompt, considering always the appearance and effect before the reality and possibilities of things. It will probably tend to form a culture about the political and financial operator as its ideal and central type, opposed to, and conflicting with, the forces of attraction that will tend to group the new social masses about the scientific engineer.* . . .

Here, then (in the vision of the present writer), are the main social elements of the coming time: (I.) the element of irresponsible property; (II.) the helpless superseded poor, that broad base of mere toilers now no longer essential; (III.) a great inchoate mass of more or less capable people engaged more or less consciously in applying the growing body of scientific knowledge to the general needs, a great mass that will inevitably tend to organise itself in a system of interdependent educated classes with a common consciousness and aim, but which may or may not succeed in doing do; and (IV.) a possibly equally great number of non-productive persons living in and by the social confusion.

All these elements will be mingled confusedly together, passing into one another by insensible gradations, scattered over the great urban regions and intervening areas our previous anticipations have sketched out. Moreover, they are developing, as it were unconsciously, under the stimulus of mechanical developments, and with the bandages of old tradition hampering their movements. The laws they obey, the governments they live under, are for the most part laws made and governments planned

*The future of the servant class and the future of the artist are two interesting questions that will be most conveniently mentioned at a later stage, when we come to discuss the domestic life in greater detail than is possible before we have formed any clear notion of the sort of people who will lead that life.

before the coming of steam. The areas of administration are still areas marked out by conditions of locomotion as obsolete as the quadrupedal method of the prearboreal ancestor. In Great Britain, for example, the political constitution, the balance of estates and the balance of parties, preserves the compromise of long-vanished antagonisms. The House of Lords is a collection of obsolete territorial dignitaries fitfully reinforced by the bishops and a miscellany (in no sense representative) of opulent moderns; the House of Commons is the seat of a party conflict, a faction fight of initiated persons, that has long ceased to bear any real relation to current social processes. The members of the lower chamber are selected by obscure party machines operating upon constituencies almost all of which have long since become too vast and heterogeneous to possess any collective intelligence of purpose at all. In theory the House of Commons guards the interests of classes that are, in fact, rapidly disintegrating into a number of quite antagonistic and conflicting elements. The new mass of capable men, of which the engineers are typical, these capable men who must necessarily be the active principle of the new mechanically equipped social body, finds no representation save by accident in either assembly. The man who has concerned himself with the public health, with army organisation, with educational improvement, or with the vital matters of transport and communication, if he enter the official councils of the kingdom at all, must enter ostensibly as the guardian of the interests of the free and independent electors of a specific district that has long ceased to have any sort of specific interests at all.* . . .

*Even the physical conditions under which the House of Commons meets and plays at government, are ridiculously obsolete. Every disputable point is settled by a division, a bell rings, there is shouting and running, the members come blundering into the chamber and sort themselves with much loutish shuffling and shoving into the division-lobbies. They are counted, as illiterate farmers count sheep; amidst much fuss and confusion they return to their places, and the tellers vociferate the result. The waste of time over these antics is enormous, and they are often repeated many times in an evening. For the lack of time, the House of Commons is unable to perform the most urgent and necessary legislative duties–it has this year hung up a cryingly necessary Education Bill, a delay that will in the end cost Great Britain millions–but not a soul in it has had the necessary commonsense to point out that an electrician and an expert locksmith could in a few weeks, and for a few hundred pounds, devise and construct a member's desk and key, committee-room tapes and voting-desks, and a general recording apparatus, that would enable every member within the precincts to vote, and that would count, record, and report the votes within the space of a couple of minutes.

And the same obsolescence that is so conspicuous in the general institutions of the official kingdom of England, and that even English people can remark in the official empire of China, is to be traced in a greater or lesser degree in the nominal organisation and public tradition throughout the whole world. The United States, for example, the social mass which has perhaps advanced furthest along the new lines, struggles in the iron bonds of a constitution that is based primarily on a conception of a number of comparatively small, internally homogenous, agricultural states, a bunch of pre-Johannesburg Transvaals, communicating little, and each constituting a separate autonomous democracy of free farmers–slaveholding or slaveless. Every country in the world, indeed, that is organised at all, has been organised with a view to stability within territorial limits; no country has been organised with any foresight of development and inevitable change, or with the slightest reference to the practical revolution in topography that the new means of transit involve. And since this is so, and since humanity is most assuredly embarked upon a series of changes of which we know as yet only the opening phases, a large part of the history of the coming years will certainly record more or less conscious endeavours to adapt these obsolete and obsolescent contrivances for the management of public affairs to the new and continually expanding and changing requirements of the social body, to correct or overcome the traditions that were once wisdom and which are now obstruction, and to burst the straining boundaries that were sufficient for the ancient states. There are here no signs of a millennium. Internal reconstruction, while men are still limited, egotistical, passionate, ignorant, and ignorantly led, means seditions and revolutions, and the rectification of frontiers means wars. But before we go on to these conflicts and wars certain general social reactions must be considered.

Let me for this new edition add a footnote to this chapter. I speak of a class of *educated and intelligent efficients* (p. 49). By that it ought to be obvious that I do not mean *trained specialists,* though a number of readers and critics have skipped to that misleading conclusion.–H. G. W.

IV. CERTAIN SOCIAL REACTIONS

WE are now in a position to point out and consider certain general ways in which the various factors and elements in the deliquescent society of the present time will react one upon another, and to speculate what definite statements, if any, it may seem reasonable to make about the individual people of the year 2000–or thereabouts–from the reaction of these classes we have attempted to define.

To begin with, it may prove convenient to speculate upon the trend of development of that class about which we have the most grounds for certainty in the coming time. The shareholding class, the rout of the Abyss, the speculator, may develop in countless ways according to the varying development of exterior influences upon them, but of the most typical portion of the central body, the section containing the scientific engineering or scientific medical sort of people, we can postulate certain tendencies with some confidence. Certain ways of thought they must develop, certain habits of mind and eye they will radiate out into the adjacent portions of the social mass. We can even, I think, deduce some conception of the home in which a fairly typical example of this body will be living within a reasonable term of years.

The mere fact that a man is an engineer or a doctor, for example, should imply now, and certainly will imply in the future, that he has received an education of a certain definite type; he will have a general acquaintance with the scientific interpretation of the universe, and he will have acquired certain positive and practical habits of mind. If the methods of thought of any individual in this central body are not practical and positive, he will tend to drift out of it to some more congenial employment. He will almost necessarily have a strong imperative to duty quite apart from whatever theological opinions he may entertain, because if he has not such an inherent imperative, life will have very many more alluring prospects than this. His religious conclusions, whatever they may be, will be based upon some orderly theological system that must have honestly admitted and reconciled his scientific beliefs; the emotional and mystical elements in his religion will be subordinate or absent. Essentially he will be a moral man, certainly so far as to exercise self-restraint and live in an ordered way. Unless this is so, he will be unable to give his principal energies to thought and work–that is, he will not be a good typical engineer. If sensuality appear at all largely in this central body, therefore,–a point we must leave open here–it will appear without any trappings of sentiment or mysticism, frankly on Pauline lines, wine for the stomach's sake, and it is better to marry than to burn, a concession to the flesh necessary to secure efficiency. Assuming in our typical case that pure indulgence does not appear or flares and passes, then either he will be single or more or less married. The import of that "more or less" will be discussed later, for the present we may very conveniently conceive him married under the traditional laws of Christendom. Having a mind considerably engaged, he will not have the leisure for a wife of the distracting, perplexing personality kind, and in our typical case, which will be a typically sound and successful one, we may picture him wedded to a healthy, intelligent, and loyal person, who will be her husband's companion in their common leisure, and as mother of their three or four children and manager of his household, as much of a technically capable individual as himself. He will be a father of several children, I think, because his scientific mental basis will incline him to see the whole of life as a struggle to survive; he will recognise that a childless, sterile life, however pleasant, is essentially failure and perversion, and he will conceive his honour involved in the possession of offspring.

Such a couple will probably dress with a view to decent convenience, they will not set the fashions, as I shall presently point out, but they will incline to steady and sober them, they will avoid exciting colour contrasts and bizarre contours. They will not be habitually promenaders, or greatly addicted to theatrical performances; they will probably find their secondary interests–the cardinal one will of course be the work in hand–in a not too imaginative prose literature, in travel and journeys and in the less sensuous aspects of music. They will probably take a considerable interest in public affairs. Their *ménage*, which will consist of father, mother, and children, will, I think, in all probability, be servantless.

They will probably not keep a servant for two very excellent reasons, because in the first place they will not want one, and in the second they will not get one if they do. A servant is necessary in the small, modern house, partly to supplement the deficiencies of the wife, but mainly to supplement the deficiencies of the house. She comes to cook and perform various skilled duties that the wife lacks either knowledge or training, or both, to perform regularly and expeditiously. Usually it must be confessed that the servant in the small household fails to perform these skilled duties completely. But the great proportion of the servant's duties consists merely in drudgery that the stupidities of our present-day method of house construction entail, and which the more sanely constructed house of the future will avoid. Consider, for instance, the wanton disregard of avoidable toil displayed in building houses with a service basement without lifts! Then most dusting and sweeping would be quite avoidable if houses were wiselier done. It is the lack of proper warming appliances which necessitates a vast amount of coal carrying and dirt distribution, and it is this dirt mainly that has so painfully to be removed again. The house of the future will probably be warmed in its walls from some power-generating station, as, indeed, already very many houses are lit at the present day. The lack of sane methods of ventilation also enhances the general dirtiness and dustiness of the present-day home, and gas-lighting and the use of tarnishable metals, wherever possible, involve further labour. But air will enter the house of the future through proper tubes in the walls, which will warm it and capture its dust, and it will be spun out again by a simple mechanism. And by simple devices such sweeping as still remains necessary can be enormously lightened. The fact that in existing homes the skirting meets the floor at right angles makes sweeping about twice as troublesome as it will be when people have the sense and ability to round off the angle between wall and floor.

So one great lump of the servant's toil will practically disappear. Two others are already disappearing. In many houses there are still the offensive duties of filling lamps and blacking boots to be done. Our coming house, however, will have no lamps to need filling, and, as for the boots, really intelligent people will feel the essential ugliness of wearing the evidence of constant manual toil upon their persons. They will wear sorts of shoes and boots that can be cleaned by wiping in a minute or so. Take now the bedroom work. The lack of ingenuity in sanitary fittings at present forbids the obvious convenience of hot and cold water supply to the bedroom, and there is a mighty fetching and carrying of water and slops to be got through daily. All that will cease. Every bedroom will have its own bath-dressing room which any well-bred person will be intelligent and considerate enough to use and leave without the slightest disarrangement. This, so far as "upstairs" goes, really only leaves bedmaking to be done, and a bed does not take five minutes to make. Downstairs a vast amount of needless labour at present arises out of table wear. "Washing up" consists of a tedious cleansing and wiping of each table utensil in turn, whereas it should be possible to immerse all dirty table wear in a suitable solvent for a few minutes and then run that off for the articles to dry. The application of solvents to window-cleaning, also, would be a possible thing but for the primitive construction of our windows, which prevents anything but a painful rub, rub, rub, with the leather. A friend of mine in domestic service tells me that this rubbing is to get the window dry, and this seems to be the general impression, but I think it incorrect. The water is not an adequate solvent, and enough cannot be used under existing conditions. Consequently, if the window is cleaned and left wet, it dries in drops, and these drops contain dirt in solution which remain as spots. But water containing a suitable solvent could quite simply be made to run down a window for a few minutes from pinholes in a pipe above into a groove below, and this could be followed by pure rain water for an equal time, and in this way the whole window cleaning in the house could, I imagine, be reduced to the business of turning on a tap.

There remains the cooking. To-day cooking, with its incidentals, is a very serious business; the coaling, the ashes, the horrible moments of heat, the hot, black things to handle, the silly, vague recipes, the want of neat apparatus, and the want of intelligence to demand or use neat apparatus. One always imagines a cook working with a crimsoned face and bare, blackened arms. But with a neat little range, heated by electricity and provided with thermometers, with absolutely controllable temperatures and proper heat screens, cooking might very easily be made a pleasant amusement for intelligent invalid ladies. Which reminds me, by-the-bye, as an added detail to our previous sketch of the scenery of the days to come, that there will be no chimneys at all to the house of the future of this type, except the flue for the kitchen smells.* This will not only abolish the chimney-stack, but make the roof a clean and pleasant addition to the garden spaces of the home.

I do not know how long all these things will take to arrive. The erection of a series of experimental labour-saving houses by some philanthropic person, for exhibition and discussion, would certainly bring about a very extraordinary advance in domestic comfort even in the immediate future, but the fashions in philanthropy do not trend in such practical directions; if they did, the philanthropic person would probably be too amenable to flattery to escape the pushful patentee and too sensitive to avail himself of criticism (which rarely succeeds in being both penetrating and polite), and it will probably be many years before the cautious enterprise of advertising firms approximates to the economies that are theoretically possible to-day. But certainly the engineering and medical sorts of person will be best able to appreciate the possibilities of cutting down the irksome labours of the contemporary home, and most likely to first demand and secure them.

The wife of this ideal home may probably have a certain distaste for vicarious labour, that so far as the immediate minimum of duties goes will probably carry her through them. There will be few servants obtainable for the small homes of the future, and that may strengthen her sentiments. Hardly any woman seems to object to a system of things which provides that another woman should be made rough-handed and kept rough-minded for her sake, but with the enormous diffusion of levelling information that is going on, a perfectly valid objection will probably come from the other side in this transaction. The servants of the past and the only good servants of to-day are the children of servants or the children of the old labour base of the social pyramid, until recently a necessary and self-respecting element in the State. Machinery has smashed that base and scattered its fragments; the tradition of self-respecting inferiority is being utterly destroyed in the world. The contingents of the Abyss, even, will not supply daughters for this purpose. In the community of the United States no native-born race of white servants has appeared, and the emancipated young negress degenerates towards the impossible–which is one of the many stimulants to small ingenuities that may help very powerfully to give that

*That interesting book by Mr. George Sutherland, *Twentieth Century Inventions,* is very suggestive on these as on many other matters.

nation the industrial leadership of the world. The servant of the future, if indeed she should still linger in the small household, will be a person alive to a social injustice and the unsuccessful rival of the wife. Such servants as wealth will retain will be about as really loyal and servile as hotel waiters, and on the same terms. For the middling sort of people in the future maintaining a separate *ménage* there is nothing for it but the practically automatic house or flat, supplemented, perhaps, by the restaurant or the hotel.

Almost certainly, for reasons detailed in the second chapter of these Anticipations, this household, if it is an ideal type, will be situated away from the central "Town" nucleus and in pleasant surroundings. And I imagine that the sort of woman who would be mother and mistress of such a home would not be perfectly content unless there were a garden about the house. On account of the servant difficulty, again, this garden would probably be less laboriously neat than many of our gardens to-day–no "bedding-out," for example, and a certain parsimony of mown lawn. . . .

To such a type of home it seems the active, scientifically trained people will tend. But usually, I think, the prophet is inclined to overestimate the number of people who will reach this condition of affairs in a generation or so, and to underestimate the conflicting tendencies that will make its attainment difficult to all, and impossible to many, and that will for many years tint and blotch the achievement of those who succeed with patches of unsympathetic colour. To understand just how modifications may come in, it is necessary to consider the probable line of development of another of the four main elements in the social body of the coming time. As a consequence and visible expression of the great new growth of share and stock property there will be scattered through the whole social body, concentrated here perhaps, and diffused there, but everywhere perceived, the members of that new class of the irresponsible wealthy, a class, as I have already pointed out in the preceding chapter, miscellaneous and free to a degree quite unprecedented in the world's history. Quite inevitably great sections of this miscellany will develop characteristics almost diametrically opposed to those of the typical working expert class, and their gravitational attraction may influence the lives of this more efficient, finally more powerful, but at present much less wealthy, class to a very considerable degree of intimacy.

The rich shareholder and the skilled expert must necessarily be sharply contrasted types, and of the two it must be borne in mind that it is the rich shareholder who spends the money.

While occupation and skill incline one towards severity and economy, leisure and unlimited means involve relaxation and demand the adventitious interest of decoration. The shareholder will be the decorative influence in the State. So far as there will be a typical shareholder's house, we may hazard that it will have rich colours, elaborate hangings, stained glass adornments, and added interests in great abundance. This "leisure class" will certainly employ the greater proportion of the artists, decorators, fabric-makers, and the like, of the coming time. It will dominate the world of art–and we may say, with some confidence, that it will influence it in certain directions. For example, standing apart from the movement of the world, as they will do to a very large extent, the archaic, opulently done, will appeal irresistibly to very many of these irresponsible rich as the very quintessence of art. They will come to art with uncritical, cultured minds, full of past achievements, ignorant of present necessities. Art will be something added to life–something stuck on and richly reminiscent–not a manner pervading all real things. We may be pretty sure that very few will grasp the fact that an iron bridge or a railway-engine may be artistically done–these will not be "art" objects, but hostile novelties. And, on the other hand, we can pretty confidently foretell a spacious future and much amplification for that turgid, costly, and deliberately anti-contemporary group of styles of which William Morris and his associates have been the fortunate pioneers. And the same principles will apply to costume. A nonfunctional class of people cannot have a functional costume, the whole scheme of costume, as it will be worn by the wealthy classes in the coming years, will necessarily be of that character which is called fancy dress. Few people will trouble to discover the most convenient forms and materials, and endeavour to simplify them and reduce them to beautiful forms, while endless enterprising tradesmen will be alert for a perpetual succession of striking novelties. The women will ransack the ages for becoming and alluring anachronisms, the men will appear in the elaborate uniforms of "games," in modifications of "court" dress, in picturesque revivals of national costumes, in epidemic fashions of the most astonishing sort. . . .

Now, these people, so far as they are spenders of money, and so far as he is a spender of money, will stand to this ideal engineering sort of person, who is the vitally important citizen of a progressive scientific State, in a competitive relation. In most cases, whenever there is something that both want, one against the other, the shareholder will get it; in most cases, where it is a matter of calling the tune, the shareholder will call the tune. For example, the young architect, conscious of exceptional ability, will have more or less clearly before him the alternatives of devoting himself to the novel, intricate, and difficult business of designing cheap, simple, and mechanically convenient homes for people who will certainly not be highly remunerative, and will probably be rather acutely critical, or of perfecting himself in some period of romantic architecture, or striking out some startling and attractive novelty of manner or material which will be certain, sooner or later, to meet its congenial shareholder. Even if he hover for a time between these alternatives, he will need to be a person not only of exceptional gifts, but what is by no means a common accompaniment of exceptional gifts, exceptional strength of character, to take the former line. Consequently, for many years yet, most of the experimental buildings and novel designs, that initiate discussion and develop the general taste, will be done primarily to please the more originative shareholders and not to satisfy the demands of our engineer or doctor; and the strictly commercial builders, who will cater for all but the wealthiest engineers, scientific investigators, and business men, being unable to afford specific designs, will–amidst the disregarded curses of these more intelligent customers–still simply reproduce in a cheaper and mutilated form such examples as happen to be set. Practically, that is to say, the shareholder will buy up almost all the available architectural talent.

This modifies our conception of the outer appearance of that little house we imagined. Unless it happens to be the house of an exceptionally prosperous member of the utilitarian professions, it will lack something of the neat directness implied in our description, something of that inevitable beauty that arises out of the perfect attainment of ends–for very many years, at anyrate. It will almost certainly be tinted, it may even be saturated, with the second-hand archaic. The owner may object, but a busy man cannot stop his life-work to teach architects what they ought to know. It may be heated electrically, but it will have sham chimneys, in whose darkness, unless they are built solid, dust and filth will gather, and luckless birds and insects pass horrible last hours of ineffectual struggle. It may have automatic window-cleaning arrangements, but they will be hidden by "picturesque" mullions. The sham chimneys will, perhaps, be made to smoke genially in winter by some ingenious contrivance, there may be sham open fireplaces within, with ingle nooks about the sham glowing logs. The needlessly steep roofs will have a sham sag and sham timbered gables, and probably forced lichens will give it a sham appearance of age. Just that feeble-minded contemporary shirking of the truth of things that has given the world such stockbroker in armour affairs as the Tower Bridge and historical romance, will, I fear, worry the lucid mind in a great multitude of the homes that the opening half, at least, of this century will produce.

In quite a similar way the shareholding body will buy up all the clever and more enterprising makers and designers of clothing and adornment, he will set the fashion of almost all ornament, in bookbinding and printing and painting, for example, furnishing, and indeed of almost all things that are not primarily produced "for the million," as the phrase goes. And where that sort of thing comes in, then, so far as the trained and intelligent type of man goes, for many years yet it will be simply a case of the nether instead of the upper millstone. Just how far the influence and contagion of the shareholding mass will reach into this imaginary household of non-shareholding efficients, and just how far the influence of science and mechanism will penetrate the minds and methods of the rich, becomes really one of the most important questions with which these speculations will deal. For this argument that he will perhaps be able to buy up the architect and the tailor and the decorator and so forth is merely preliminary to the graver issue. It is just possible that the shareholder may, to a very large extent–in a certain figurative sense, at least–buy up much of the womankind that would otherwise be available to constitute those severe, capable, and probably by no means unhappy little establishments to which our typical engineers will tend, and so prevent many women from becoming mothers of a regenerating world. The huge secretion of irresponsible wealth by the social organism is certain to affect the tone of thought of the entire feminine sex profoundly–the exact nature of this influence we may now consider.

The gist of this inquiry lies in the fact that, while a man's starting position in this world of to-day is entirely determined by the conditions of his birth and early training, and his final position the slow elaborate outcome of his own sustained efforts to live, a woman, from the age of sixteen onward–as the world goes now–is essentially adventurous, the creature of circumstances largely beyond her control and foresight. A virile man, though he, too, is subject to accidents, may, upon most points, still hope to plan and determine his life; the life of a woman is all accident. Normally she lives in relation to some specific man, and until that man is indicated her preparation for life must be of the most tentative sort. She lives, going nowhere, like a cabman on the crawl, and at any time she may find it open to her to assist some pleasure-loving millionaire to spend his millions, or to play her part in one of the many real, original, and only derivatives of the former aristocratic "Society" that have developed themselves among independent people. Even if she is a serious and labour-loving type, some shareholder may tempt her with the prospect of developing her exceptional personality in ease and freedom and in "doing good" with his money. With the continued growth of the shareholding class, the brighter-looking matrimonial chances, not to speak of the glittering opportunities that are not matrimonial, will increase. Reading is now the privilege of all classes, there are few secrets of etiquette that a clever lower-class girl will fail to learn, there are few such girls, even now, who are not aware of their wide opportunities, or at least their wide possibilities, of luxury and freedom, there are still fewer who, knowing as much, do not let it affect their standards and conception of life. The whole mass of modern fiction written by women for women, indeed, down to the cheapest novelettes, is saturated with the romance of *mésalliance*. And even when the specific man has appeared, the adventurous is still not shut out of a woman's career. A man's affections may wander capriciously and leave him but a little poorer or a little better placed; for the women they wander from, however, the issue is an infinitely graver one, and the serious wandering of a woman's fancy may mean the beginning of a new world for her. At any moment the chances of death may make the wife a widow, may sweep out of existence all that she had made fundamental in her life, may enrich her with insurance profits or hurl her into poverty, and restore all the drifting expectancy of her adolescence. . . .

Now, it is difficult to say why we should expect the growing girl, in whom an unlimited ambition and egotism is as natural and proper a thing as beauty and high spirits, to deny herself some dalliance with the more opulent dreams that form the golden lining to these precarious prospects? How can we expect her to prepare herself solely, putting all wandering thoughts aside, for the servantless cookery, domestic *Kindergarten* work, the care of hardy perennials, and low-pitched conversation of the engineer's home? Supposing, after all, there is no predestinate engineer! The stories the growing girl now prefers, and I imagine will in the future still prefer, deal mainly with the rich and free; the theatre she will prefer to visit will present the lives and loves of opulent people with great precision and detailed correctness; her favourite periodicals will reflect that life; her schoolmistress, whatever her principles, must have an eye to her "chances." And even after Fate or a gust of passion has whirled her into the arms of our busy and capable fundamental man, all these things will still be in her imagination and memory. Unless he is a person of extraordinary mental prepotency, she will almost insensibly determine the character of the home in a direction quite other than that of our first sketch. She will set herself to realise, as far as her husband's means and credit permit, the ideas of the particular section of the wealthy that have captured her. If she is a fool, her ideas of life will presently come into complete conflict with her husband's in a manner that, as the fumes of the love-potion leave his brain, may bring the real nature of the case home to him. If he is of that resolute strain to whom the world must finally come, he may rebel and wade through tears and crises to his appointed work again. The cleverer she is, and the finer and more loyal her character up to a certain point, the less likely this is to happen, the more subtle and effective will be her hold upon her husband, and the more probable his perversion from the austere pursuit of some interesting employment, towards the adventures of modern money-getting in pursuit of her ideals of a befitting life. And meanwhile, since "one must live," the nursery that was implicit in the background of the first picture will probably prove unnecessary. She will be, perforce, a person not only of pleasant pursuits, but of leisure. If she endears herself to her husband, he will feel not only the attraction but the duty of her vacant hours; he will not only deflect his working hours from the effective to the profitable, but that occasional burning of the midnight oil, that no brainworker may forego if he is to retain his efficiency, will, in the interests of some attractive theatrical performance or some agreeable social occasion, all too frequently have to be put off or abandoned.

This line of speculation, therefore, gives us a second picture of a household to put beside our first, a household, or rather a couple, rather more likely to be typical of the mass of middling sort of people in those urban regions of the future than our first projection. It will probably not live in a separate home at all, but in a flat in "Town," or at one of the subordinate centres of the urban region we have foreseen. The apartments will be more or less agreeably adorned in some decorative fashion akin to but less costly than some of the many fashions that will obtain among the wealthy. They will be littered with a miscellaneous literature, novels of an entertaining and stimulating sort predominating, and with *bric-à-brac;* in a childless household, there must certainly be quaint dolls, pet images, and so forth, and perhaps a canary would find a place. I suspect there would be an edition or so of "Omar" about in this more typical household of "Moderns," but I doubt about the Bible. The man's working books would probably be shabby and relegated to a small study, and even these overlaid by abundant copies of the *Financial*-something or other. It would still be a servantless household, and probably not only without a nursery but without a kitchen, and in its grade and degree it would probably have social relations directly or intermediately through rich friends with some section, some one of the numerous cults of the quite independent wealthy.

Quite similar households to this would be even more common among those neither independent nor engaged in work of a primarily functional nature, but endeavouring quite ostensibly to acquire wealth by political or business ingenuity and activity, and also among the great multitude of artists, writers, and that sort of people, whose works are their children. In comparison with the state of affairs fifty years ago, the child-infested household is already conspicuously rare in these classes.

These are two highly probably *ménages* among the central mass of the people of the coming time. But there will be many others. The *ménage à deux,* one may remark, though it may be without the presence of children, is not necessarily childless. Parentage is certainly part of the pride of many men–though, curiously enough, it does not appear to be felt among modern European married women as any part of their honour. Many men will probably achieve parentage, therefore, who will not succeed in inducing, or who may possibly even be very loth to permit, their wives to undertake more than the first beginnings of motherhood. From the moment of its birth, unless it is kept as a pet, the child of such marriages will be nourished, taught, and trained almost as though it were an orphan, it will have a succession of bottles and foster-mothers for body and mind from the very beginning. Side by side with this increasing number of childless homes, therefore, there may develop a system of *Kindergarten* boarding-schools. Indeed, to a certain extent such schools already exist, and it is one of the unperceived contrasts of this and any former time how common such a separation of parents and children becomes. Except in the case of the illegitimate and orphans, and the children of impossible (many public-house children, *e.g.*), or wretched homes, boarding-schools until quite recently were used only for quite big boys and girls. But now, at every seaside town, for example, one sees a multitude of preparatory schools, which are really not simply educational institutions, but supplementary homes. In many cases these are conducted and very largely staffed by unmarried girls and women who are indeed, in effect, assistant mothers. This class of capable schoolmistresses is one of the most interesting social developments of this period. For the most part they are women who from emotional fastidiousness, intellectual egotism, or an honest lack of passion, have refused the common lot of marriage, women often of exceptional character and restraint, and it is well that, at anyrate, their intelligence and character should not pass fruitlessly out of being. Assuredly for this type the future has much in store.

There are, however, still other possibilities to be considered in this matter. In these Anticipations it is impossible to ignore the forces making for a considerable relaxation of the institution of permanent monogamous marriage on the coming years, and of a much greater variety of establishments than is suggested by these possibilities within the pale. I guess, without attempting to refer to statistics, that our present society must show a quite unprecedented number and increasing number of male and female celibates–not religious celibates, but people, for the most part, whose standard of personal comfort has such a relation to their earning power that they shirk or cannot enter the matrimonial grouping. The institution of permanent monogamous marriage–except in the ideal Roman Catholic community, where it is based on the sanction of an authority which in Roman Catholic countries a large proportion of the men decline to obey–is sustained at present entirely by the inertia of custom, and by a number of sentimental and practical considerations, considerations that may very possibly undergo modification in the face of the altered relationship of husband and wife that the present development of childless *ménages* is bringing about. The practical and sustaining reason for monogamy is the stability it gives to the family; the value of a stable family lies in the orderly upbringing in an atmosphere of affection that it secures in most cases for its more or less numerous children. The monogamous family has indisputably been the civilising unit of the pre-mechanical civilised state. It must be remembered that both for husband and wife in most cases monogamic life marriage involves an element of sacrifice, it is an institution of late appearance in the history of mankind, and it does not completely fit the psychology or physiology of any but very exceptional characters in either sex. For the man it commonly involves considerable restraint; he must ride his imagination on the curb, or exceed the code in an extremely dishonouring, furtive, and unsatisfactory manner while publicly professing an impossible virtue; for the woman it commonly implies many uncongenial submissions. There are probably few married couples who have escaped distressful phases of bitterness and tears, within the constraint of their, in most cases, practically insoluble bond. But, on the other hand, and as a reward that in the soberer, mainly agricultural civilisation of the past, and among the middling class of people, at anyrate, has sufficed, there comes the great development of associations and tendernesses that arises out of intimate co-operation in an established home, and particularly out of the linking love and interest of children's lives. . . .

But how does this fit into the childless, disunited, and probably shifting *ménage* of our second picture?

It must be borne in mind that it has been the middling and lower mass of people, the tenants and agriculturists, the shopkeepers, and so forth, men needing before all things the absolutely loyal help of wives, that has sustained permanent monogamic marriage whenever it has been sustained. Public monogamy has existed on its merits–that is, on the merits of the wife. Merely ostensible reasons have never sufficed. No sort of religious conviction, without a real practical utility, has ever availed to keep classes of men, unhampered by circumstances, to its restrictions. In all times, and holding all sorts of beliefs, the specimen humanity of courts and nobilities is to be found developing the most complex qualifications of the code. In some quiet corner of Elysium the bishops of the early Georges, the ecclesiastical dignitaries of the contemporary French and Spanish courts, the patriarchs of vanished Byzantium, will find a common topic with the spiritual advisers of the kingdoms of the East in this difficult theme,–the theme of the concessions permissible and expedient to earnest believers encumbered with leisure and a superfluity of power. . . . It is not necessary to discuss religious development, therefore, before deciding this issue. We are dealing now with things deeper and forces infinitely more powerful than the mere convictions of men.

Will a generation to whom marriage will be no longer necessarily associated with the birth and rearing of children, or with the immediate co-operation and sympathy of husband and wife in common proceedings, retain its present feeling for the extreme sanctity of the permanent bond? Will the agreeable, unemployed, childless woman, with a high conception of her personal rights, who is spending her husband's earnings or income in some pleasant discrepant manner, a type of woman there are excellent reasons for anticipating will become more frequent–will she continue to share the honours and privileges of the wife, mother, and helper of the old dispensation? and in particular, will the great gulf that is now fixed by custom between her and the agreeable unmarried lady who is similarly employed remain so inexorably wide? Charity is in the air, and why should not charming people meet one another? And where is either of these ladies to find the support that will enable her to insist upon the monopoly that conventional sentiment, so far as it finds expression, concedes her? The danger to them both of the theory of equal liberty is evident enough. On the other hand, in the case of the unmarried mother who may be helped to hold her own, or who may be holding her own in the world, where will the moral censor of the year 1950 find his congenial following to gather stones? Much as we may regret it, it does very greatly affect the realities of this matter, that with the increased migration of people from home to home amidst the large urban regions that, we have concluded, will certainly obtain in the future, even if moral reprobation and minor social inconveniences do still attach to certain sorts of status, it will probably be increasingly difficult to determine the status of people who wish to conceal it for any but criminal ends.

In another direction there must be a movement towards the relaxation of the marriage law and of divorce that will complicate status very confusingly. In the past it has been possible to sustain several contrasting moral systems in each of the practically autonomous states of the world, but with a development and cheapening of travel and migration that is as yet only in its opening phase, an increasing conflict between dissimilar moral restrictions must appear. Even at present, with only the most prosperous classes of the American and Western European countries migrating at all freely, there is a growing amount of inconvenience arising out of these–from the point of view of social physiology–quite arbitrary differences. A man or woman may, for example, have been the injured party in some conjugal complication, may have established a domicile and divorced the erring spouse in certain of the United States, may have married again there with absolute local propriety, and may be a bigamist and a criminal in England. A child may be a legal child in Denmark or Australia, and a bastard in this austerer climate. These things are, however, only the first intimations of much more profound reactions. Almost all the great European Powers, and the United States also, are extending their boundaries to include great masses of non-Christian polygamous peoples, and they are permeating these peoples with railways, printed matter, and all the stimulants of our present state. With the spread of these conveniences there is no corresponding spread of Christianity. These people will not always remain in the ring fence of their present regions; their superseded princes, and rulers, and public masters, and managers, will presently come to swell the shareholding mass of the appropriating Empire. Europeans, on the other hand, will drift into these districts, and under the influence of their customs, intermarriages and interracial reaction will increase; in a world which is steadily abolishing locality, the compromise of local concessions, of localised recognition of the "custom of the country," cannot permanently avail. Statesmen will have to face the alternative of either widening the permissible variations of the marriage contract, or of acute racial and religious stresses, of a vast variety of possible legal betrayals, and the appearance of a body of self-respecting people, outside the law and public respect, a body that will confer a touch of credit upon, because it will share the stigma of, the deliberately dissolute and criminal. And whether the moral law shrivels relatively by mere exclusiveness—as in religious matters the Church of England, for example, has shrivelled to the proportions of a mere sectarian practice—or whether it broadens itself to sustain justice in a variety of sexual contracts, the net result, so far as our present purpose goes, will be the same. All these forces, making for moral relaxation in the coming time, will probably be greatly enhanced by the line of development certain sections of the irresponsible wealthy will almost certainly follow.

Let me repeat that the shareholding rich man of the new time is in a position of freedom almost unparalleled in the history of men. He has sold his permission to control and experiment with the material wealth of the community for freedom—for freedom from care, labour, responsibility, custom, local usage and local attachment. He may come back again into public affairs if he likes—that is his private concern. Within the limits of the law and his capacity and courage, he may do as the imagination of his heart directs. Now, such an experimental and imperfect creature as man, a creature urged by such imperious passions, so weak in imagination and controlled by so feeble a reason, receives such absolute freedom as this only at infinite peril. To a great number of these people, in the second or third generation, this freedom will mean vice, the subversion of passion to inconsequent pleasures. We have on record, in the personal history of the Roman emperors, how freedom and uncontrolled power took one representative group of men, men not entirely of one blood nor of one bias, but reinforced by the arbitrary caprice of adoption and political revolution. We have in the history of the Russian empresses a glimpse of similar feminine possibilities. We are moving towards a time when, through this confusion of moral standards I have foretold, the pressure of public opinion in these matters must be greatly relaxed, when religion will no longer speak with a unanimous voice, and when freedom of escape from disapproving neighbours will be greatly facilitated. In the past, when depravity had a centre about a court, the contagion of its example was limited to the court region, but every idle rich man of this great, various, and widely diffused class, will play to a certain extent the moral *rôle* of a court. In these days of universal reading and vivid journalism, every novel infraction of the code will be known of, thought about, and more or less thoroughly discussed by an enormous and increasing proportion of the common people. In the past it has been possible for the churches to maintain an attitude of respectful regret towards the lapses of the great, and even to co-operate in these lapses with a sympathetic privacy, while maintaining a wholesome rigour towards vulgar vice. But in the coming time there will be no Great, but many rich, the middling sort of people will probably be better educated as a whole than the rich, and the days of their differential treatment are at an end.

It is foolish, in view of all these things, not to anticipate and prepare for a state of things when not only will moral standards be shifting and uncertain, admitting of physiologically sound *ménages* of very variable status, but also when vice and depravity, in every form that is not absolutely penal, will be practised in every grade of magnificence and condoned. This means that not only will status cease to be simple and become complex and varied, but that outside the system of *ménages* now recognised, and under the disguise of which all other *ménages* shelter, there will be a vast drifting and unstable population grouped in almost every conceivable form of relation. The world of Georgian England was a world of Homes; the world of the coming time will still have its Homes, its real Mothers, the custodians of the human succession, and its cared-for children, the inheritors of the future, but in addition to this Home world, frothing tumultuously over and amidst these stable rocks, there will be an enormous complex of establishments, and hotels, and sterile households, and flats, and all the elaborate furnishing and appliances of a luxurious extinction.

And since in the present social chaos there does not yet exist any considerable body of citizens–comparable to the agricultural and commercial middle class of England during the period of limited monarchy–that will be practically unanimous in upholding any body of rules of moral restraint, since there will probably not appear for some generations anybody propounding with wide-reaching authority a new definitely different code to replace the one that is now likely to be increasingly disregarded, it follows that the present code with a few interlined qualifications and grudging legal concessions will remain nominally operative in sentiment and practice while being practically disregarded, glossed, or replaced in numberless directions. It must be pointed out that in effect, what is here forecast for questions of *ménage* and moral restraints has already happened to a very large extent in religious matters. There was a time when it was held–and I think rightly–that a man's religious beliefs, and particularly his method of expressing them, was a part not of his individual but of his social life. But the great upheavals of the Reformation resulted finally in a compromise, a sort of truce, that has put religious belief very largely out of intercourse and discussion. It is conceded that within the bounds of the general peace and security a man may believe and express his belief in matters of religion as he pleases, not because it is better so, but because for the present epoch there is no way nor hope of attaining unanimous truth. There is a decided tendency that will, I believe, prevail towards the same compromise in the question of private morals. There is a convention to avoid all discussion of creeds in general social intercourse; and a similar convention to avoid the point of status in relation to marriage, one may very reasonably anticipate, will be similarly recognised.

But this impending dissolution of a common standard of morals does not mean universal depravity until some great reconstruction obtains any more than the obsolescence of the Conventicle Act means universal irreligion. It means that for one Morality there will be many moralities. Each human being will, in the face of circumstances, work out his or her particular early training as his or her character determines. And although there will be a general convention upon which the most diverse people will meet, it will only be with persons who have come to identical or similar conclusions in the matter of moral conduct and who are living in similar *ménages,* just as now it is only with people whose conversation implies a certain community or kinship of religious belief, that really frequent and intimate intercourse will go on. In other words, there will be a process of moral segregation* set up. Indeed, such a process is probably already in operation, amidst the deliquescent social mass. People will be drawn together into little groups of similar *ménages* having much in common. And this–in view of the considerations advanced in the first two chapters, considerations all converging on the practical abolition of distances and the general freedom of people to live anywhere they like over large areas–will mean very frequently an actual local segregation. There will be districts that will be clearly recognised and marked as "nice," fast regions, areas of ramshackle Bohemianism, regions of earnest and active work, old-fashioned corners and Hill Tops. Whole regions will be set aside for the purposes of opulent enjoyment–a thing already happening, indeed, at points along the Riviera to-day. Already the superficial possibilities of such a segregation have been glanced at. It has been pointed out that the enormous urban region of the future may present an extraordinary variety of districts, suburbs, and subordinate centres within its limiting boundaries, and here we have a very definite enforcement of that probability.

In the previous chapter I spoke of boating centres and horsey suburbs, and picturesque hilly districts and living-places by the sea, of promenade centres and theatrical districts; I hinted at various fashions in architecture, and suchlike things, but these exterior appearances will be but the outward and visible sign of inward and more spiritual distinctions. The people who live in the good hunting country and about that glittering Grand Stand, will no longer be even pretending to live under the same code as those picturesque musical people who have concentrated on the canoe-dotted river. Where the promenaders gather, and the bands are playing, and the pretty little theatres compete, the pleasure-seeker will be seeking such pleasure as he pleases, no longer debased by furtiveness and innuendo, going his primrose path to a congenial, picturesque, happy and highly desirable extinction. Just over the hills, perhaps, a handful of opulent shareholders will be pleasantly preserving the old traditions of a landed aristocracy, with servants, tenants, vicar, and other dependents all complete, and what from the point of view of social physiology will really be an arrested contingent of the

*I use the word "segregation" here and always as it is used by mineralogists to express the slow conveyance of diffused matter upon centres of aggregation, such a process as, for example, must have occurred in the growth of flints.

Abyss, but all nicely washed and done good to, will pursue home industries in model cottages in a quite old English and exemplary manner. Here the windmills will spin and the waterfalls be trapped to gather force, and the quiet-eyed master of the machinery will have his office and perhaps his private home. Here about the great college and its big laboratories there will be men and women reasoning and studying; and here, where the homes thicken among the ripe gardens, one will hear the laughter of playing children, the singing of children in their schools, and see their little figures going to and fro amidst the trees and flowers. . . .

And these segregations, based primarily on a difference in moral ideas and pursuits and ideals, will probably round off and complete themselves at last as distinct and separate cultures. As the moral ideas realise themselves in *ménage* and habits, so the ideals will seek to find expression in a literature, and the passive drifting together will pass over into a phase of more or less conscious and intentional organisation. The segregating groups will develop fashions of costume, types of manners and bearing, and even, perhaps, be characterised by a certain type of facial expression. And this gives us a glimpse, an aspect of the immediate future of literature. The kingdoms of the past were little things, and above the mass of peasants who lived and obeyed and died, there was just one little culture to which all must needs conform. Literature was universal within the limits of its language. Where differences of view arose there were violent controversies, polemics, and persecutions, until one or other rendering had won its ascendency. But this new world into which we are passing will, for several generations at least, albeit it will be freely intercommunicating and like a whispering gallery for things outspoken, possess no universal ideals, no universal conventions: there will be the literature of the thought and effort of this sort of people, and the literature, thought, and effort of that.* Life is already

*Already this is becoming apparent enough. The literary "Boom," for example, affected the entire reading public of the early nineteenth century. It was no figure of speech that "everyone" was reading Byron or puzzling about the Waverley mystery, that first and most successful use of the unknown author dodge. The booming of Dickens, too, forced him even into the reluctant hands of Omar's Fitzgerald. But the factory-syren voice of the modern "boomster" touches whole sections of the reading public no more than fog-horns going down Channel. One would as soon think of Skinner's Soap for one's library as So-and-so's Hundred Thousand Copy Success. Instead of "everyone" talking of the Great New Book, quite considerable numbers are shamelessly admitting they don't read that sort of thing. One gets used to literary booms just as one gets used to motor cars, they are no longer marvellous, universally significant

most wonderfully arbitrary and experimental, and for the coming century this must be its essential social history, a great drifting and unrest of people, a shifting and regrouping and breaking up again of groups, great multitudes seeking to find themselves.

The safe life in the old order, where one did this because it was right, and that because it was the custom, when one shunned this and hated that, as lead runs into a mould, all that is passing away. And presently, as the new century opens out, there will become more and more distinctly emergent many new cultures and settled ways. The grey expanse of life to-day is grey, not in its essence, but because of the minute confused mingling and mutual cancelling of many-coloured lives. Presently these tints and shades will gather together here as a mass of one colour, and there as a mass of another. And as these colours intensify and the tradition of the former order fades, as these cultures become more and more shaped and conscious, as the new literatures grow in substance and power, as differences develop from speculative matter of opinion to definite intentions, as contrasts and affinities grow sharper and clearer, there must follow some very extensive modifications in the collective public life. But one series of tints, one colour must needs have a heightening value amidst this iridescent display. While the forces at work in the wealthy and purely speculative groups of society make for disintegration, and in many cases for positive elimination, the forces that bring together the really functional people will tend more and more to impose upon them certain common characteristics and beliefs, and the discovery of a group of similar and compatible class interests upon which they can unite. The practical people, the engineering and medical and scientific people, will become more and more homogeneous in their fundamental culture, more and more distinctly aware of a common "general reason" in things, and of a common difference from the less functional masses and from any sort of people in the past. They

things, but merely something that goes by with much unnecessary noise and leaves a faint offence in the air. Distinctly we segregate. And while no one dominates, while for all this bawling there are really no great authors of imperial dimensions, indeed no great successes to compare with the Waverley boom, or the boom of Macaulay's History, many men, too fine, too subtle, too aberrant, too unusually fresh for any but exceptional readers, men who would probably have failed to get a hearing at all in the past, can now subsist quite happily with the little sect they have found, or that has found them. They live safely in their islands; a little while ago they could not have lived at all, or could have lived only on the shameful bread of patronage, and yet it is these very men who are often most covetously bitter against the vulgar preferences of the present day.

will have in their positive science a common ground for understanding the real pride of life, the real reason for the incidental nastiness of vice, and so they will be a sanely reproductive class, and, above all, an educating class. Just how much they will have kept or changed of the deliquescent morality of to-day, when in a hundred years or so they do distinctively and powerfully emerge, I cannot speculate now. They will certainly be a moral people. They will have developed the literature of their needs, they will have discussed and tested and thrashed out many things, they will be clear where we are confused, resolved where we are undecided and weak. In the districts of industrial possibility, in the healthier quarters of the town regions, away from the swamps and away from the glare of the midnight lights, these people will be gathered together. They will be linked in processions through the agency of great and sober papers–in England the *Lancet,* the *British Medical Journal,* and the already great periodicals of the engineering trades, foreshadow something, but only a very little, of what these papers may be. The best of the wealthy will gravitate to their attracting centres. . . . Unless some great catastrophe in nature break down all that man has built, these great kindred groups of capable men and educated, adequate women must be, under the operation of the forces we have considered so far, the element finally emergent amidst the vast confusions of the coming time.

I take the opportunity afforded by a fresh printing of this book to add a clarifying word or so to the two preceding chapters. Much comment has made it clear to me that the shareholding and efficient elements have been too sharply antagonised in my discussion, so that it is assumed I write of two distinct, contrasted, separated strata of people. Such was certainly not my design. Throughout, my intention at least, was to contrast social forces or elements that more often than not will be found in conflict in individual men and individual households, not to contrast *classes* in the community,–much less stratified classes. For example, I do not think of the men who concentrate and control Trusts, as members of, what I will admit I have spoken of too carelessly, as the shareholding *class.* Their wives and children *may* be,–that is another matter. The essential feature of the Trust process is to gather together shares in order to control a reality; the chief feature of that diffused body I intend when I speak of the shareholding class is its parasitic detachment from reality.–H. G. W.

V. THE LIFE-HISTORY OF DEMOCRACY

IN the preceding four chapters there has been developed, in a clumsy, laborious way, a smudgy, imperfect picture of the generalised civilised state of the coming century. In terms, vague enough at times, but never absolutely indefinite, the general distribution of the population in this state has been discussed, and its natural development into four great–but in practice intimately interfused–classes. It has been shown–I know not how convincingly–that as the result of forces that are practically irresistible, a world-wide process of social and moral deliquescence is in progress, and that a really functional social body of engineering, managing men, scientifically trained, and having common ideals and interests, is likely to segregate and disentangle itself from our present confusion of aimless and ill-directed lives. It has been pointed out that life is presenting an unprecedented and increasing variety of morals, *ménages,* occupations and types, at present so mingled as to give a general effect of greyness, but containing the promise of local concentration that may presently change that greyness into kaleidoscopic effects. That image of concentrating contrasted colours will be greatly repeated in this present chapter. In the course of these inquiries, we have permitted ourselves to take a few concrete glimpses of households, costumes, conveyances, and conveniences of the coming time, but only as incidental realisations of points in this general thesis. And now, assuming, as we must necessarily do, the soundness of these earlier speculations, we have arrived at a stage when we may consider how the existing arrangements for the ostensible government of the State are likely to develop through their own inherent forces, and how they are likely to be affected by the processes we have forecast.

So far, this has been a speculation upon the probable development of a civilised society *in vacuo*. Attention has been almost exclusively given to the forces of development, and not to the forces of conflict and restraint. We have ignored the boundaries of language that are flung athwart the great lines of modern communication, we have disregarded the friction of tariffs, the peculiar groups of prejudices and irrational instincts that inspire one miscellany of shareholders, workers, financiers, and superfluous poor such as the English, to hate, exasperate, lie about, and injure another such miscellany as the French or the Germans. Moreover, we have taken very little account of the fact that, quite apart from nationality, each individual case of the new social order is developing within the form of a legal government based on conceptions of a society that has been superseded by the advent of mechanism. It is this last matter that we are about to take into consideration.

Now, this age is being constantly described as a "Democratic" age; "Democracy" is alleged to have affected art, literature, trade and religion alike in the most remarkable ways. It is not only tacitly present in the great bulk of contemporary thought that this "Democracy" is now dominant, but that it is becoming more and more overwhelmingly predominant as the years pass. Allusions to Democracy are so abundant, deductions from its influence so confident and universal, that it is worth while to point out what a very hollow thing the word in most cases really is, a large empty object in thought, of the most vague and faded associations and the most attenuated content, and to inquire just exactly what the original implications and present realities of "Democracy" may be. The inquiry will leave us with a very different conception of the nature and future of this sort of political arrangement from that generally assumed. We have already seen in the discussion of the growth of great cities, that an analytical process may absolutely invert the expectation based on the gross results up to date, and I believe it will be equally possible to show cause for believing that the development of Democracy also is, after all, not the opening phase of a worldwide movement going on unbendingly in its present direction, but the first impulse of forces that will finally sweep round into a quite different path. Flying off at a tangent is probably one of the gravest dangers and certainly the one most constantly present, in this enterprise of prophecy.

One may, I suppose, take the Rights of Man as they are embodied in the French Declaration as the ostentations of Democracy; our present Democratic state may be regarded as a practical realisation of these claims. As far as the individual goes, the realisation takes the form of an untrammelled liberty in matters that have heretofore been considered a part of social procedure, in the lifting of positive religious and moral compulsions, in the recognition of absolute property, and in the abolition of special privileges and special restrictions. Politically modern Democracy takes the form of denying that any specific person or persons shall act as a matter of intrinsic right or capacity on behalf of the community as a whole. Its root idea is representation. Government is based primarily on election, and every ruler is, in theory at least, a delegate and servant of the popular will. It is implicit in the Democratic theory that there *is* such a thing as a popular will, and this is supposed to be the net sum of the wills of all the citizens in the State, so far as public affairs are concerned. In its less perfect and more usual state the Democratic theory is advanced either as an ethical theory which postulates an absence of formal acquiescence on the part of the governed as injustice, or else as a convenient political compromise, the least objectionable of all possible methods of public control, because it will permit only the minimum of general unhappiness. . . . I know of no case for the elective Democratic government of modern States that cannot be knocked to pieces in five minutes. It is manifest that upon countless important public issues there is no collective will, and nothing in the mind of the average man except blank indifference; that an electional system simply places power in the hands of the most skilful electioneers; that neither men nor their rights are identically equal, but vary with every individual, and, above all, that the minimum or maximum of general happiness is related only so indirectly to the public control that people will suffer great miseries from their governments unresistingly, and, on the other hand, change their rulers on account of the most trivial irritations. The case against all the prolusions of ostensible Democracy is indeed so strong that it is impossible to consider the present wide establishment of Democratic institutions as being the outcome of any process of intellectual conviction; it arouses suspicion even whether ostensible Democracy may not be a mere rhetorical garment for essentially different facts, and upon that suspicion we will now inquire.

Democracy of the modern type, manhood suffrage and so forth, became a conspicuous phenomenon in the world only in the closing decades of the eighteenth century. Its genesis is so intimately connected with the first expansion of the productive element in the State, through mechanism and a co-operative organisation, as to point at once to a causative connection. The more closely one looks into the social and political life of the eighteenth century the more plausible becomes this view. New and potentially influential social factors had begun to appear—the organising manufacturer, the intelligent worker, the skilled tenant, and the urban abyss, and the traditions of the old land-owning, non-progressive aristocratic monarchy that prevailed in Christendom, rendered it incapable—without some destructive shock or convulsion—of any re-organisation to incorporate or control these new factors. In the case of the British Empire an additional stress was created by the incapacity of the formal government to assimilate the developing civilisation of the American colonies. Everywhere there were new elements, not as yet clearly analysed or defined, arising as mechanism arose; everywhere the old traditional government and social system, defined and analysed all too well, appeared increasingly obstructive, irrational, and feeble in its attempts to include and direct these new powers. But now comes a point to which I am inclined to attach very great importance. The new powers were as yet shapeless. It was not the conflict of a new organisation with the old. It was the preliminary dwarfing and deliquescence of the mature old beside the embryonic mass of the new. It was impossible then—it is, I believe, only beginning to be possible now—to estimate the proportions, possibilities, and inter-relations of the new social orders out of which a social organisation has still to be built in the coming years. No formula of definite reconstruction had been evolved, or has even been evolved yet, after a hundred years. And these swelling, inchoate new powers, whose very birth condition was the crippling, modification, or destruction of the old order, were almost forced to formulate their proceedings for a time, therefore, in general affirmative propositions that were really in effect not affirmative propositions at all, but propositions of repudiation and denial. "These kings and nobles and people privileged in relation to obsolescent functions cannot manage our affairs"—that was evident enough, that was the really essential question at that time, and since no other effectual substitute appeared ready-made, the working doctrine of the infallible judgment of humanity in the gross, as distinguished from the quite indisputable incapacity of sample individuals, became, in spite of its inherent absurdity, a convenient and acceptable working hypothesis.

Modern Democracy thus came into being, not, as eloquent persons have pretended, by the sovereign people consciously and definitely assuming power–I imagine the sovereign people in France during the first Revolution, for example, quite amazed and muddle-headed with it all–but by the decline of the old ruling classes in the face of the *quasi*-natural growth of mechanism and industrialism, and by the unpreparedness and want of organisation in the new intelligent elements in the State. I have compared the human beings in society to a great increasing variety of colours tumultuously smashed up together, and giving at present a general and quite illusory effect of grey, and I have attempted to show that there is a process in progress that will amount at last to the segregation of these mingled tints into recognisable distinct masses again. It is not a monotony, but an utterly disorderly and confusing variety that makes this grey, but Democracy, for practical purposes, does really assume such a monotony. Like ∞, the Democratic formula is a concrete-looking and negotiable symbol for a negation. It is the aspect in political disputes and contrivances of that social and moral deliquescence the nature and possibilities of which have been discussed in the preceding chapters of this volume.

Modern Democracy first asserted itself in the ancient kingdoms of France and Great Britain (counting the former British colonies in America as a part of the latter), and it is in the French and English-speaking communities that Democracy has developed itself most completely. Upon the supposition we have made, Democracy broke out first in these States because they were leading the way in material progress, because they were the first States to develop industrialism, wholesale mechanisms, and great masses of insubordinate activity outside the recognised political scheme, and the nature and time and violence of the outbreak was determined by the nature of the superseded government, and the amount of stress between it and the new elements. But the detachment of a great section of the new middle class from the aristocratic order of England to form the United States of America, and the sudden rejuvenescence of France by the swift and thorough sloughing of its outworn aristocratic monarchy, the consequent wars and the Napoleonic adventure, checked and modified the parallel development that might otherwise have happened in country after country over all Europe west of the Carpathians. The monarchies that would probably have collapsed through internal forces and given place to modern democratic states were smashed from the outside, and a process of political reconstruction, that has probably missed out the complete formal Democratic phase altogether–and which has been enormously complicated through religious, national, and dynastic traditions–set in. Throughout America, in England, and, after extraordinary experiments, in France, political democracy has in effect legally established itself–most completely in the United States–and the reflection and influence of its methods upon the methods of all the other countries in intellectual contact with it, have been so considerable as practically to make their monarchies as new in their kind, almost, as democratic republics. In Germany, Austria, and Italy, for example, there is a press nearly as audible as in the more frankly democratic countries, and measurably akin in influence; there are constitutionally established legislative assemblies, and there is the same unofficial development of powerful financial and industrial powers with which the ostensible Government must make terms. In a vast amount of the public discussion of these States, the postulates of Democracy are clearly implicit. Quite as much in reality as the democratic republics of America, are they based not on classes but upon a confusion; they are, in their various degrees and with their various individual differences, just as truly governments of the grey.

It has been argued that the grey is illusory and must sooner or later pass, and that the colour that will emerge to predominance will take its shape as a scientifically trained middle class of an unprecedented sort, not arising out of the older middle classes, but replacing them. This class will become, I believe, at last consciously *the* State, controlling and restricting very greatly the three non-functional masses with which it is as yet almost indistinguishably mingled. The general nature of its formation within the existing confusion and its emergence may, I think, with a certain degree of confidence, be already forecast, albeit at present its beginnings are singularly unpromising and faint. At present the class of specially trained and capable people–doctors, engineers, scientific men of all sorts–is quite disproportionally absent from political life, it does not exist as a factor in that life, it is growing up outside that life, and has still to develop, much more to display, a collective intention to come specifically in. But the forces are in active operation to drag it into the centre of the stage for all that.

The modern democracy or democratic quasi-monarchy conducts its affairs as though there was no such thing as special knowledge or practical education. The utmost recognition it affords to the man who has taken the pains to know, and specifically to do, is occasionally to consult him upon specific points and override his counsels in its ampler wisdom, or to entrust to him some otherwise impossible duty under circumstances of extreme limitation. The man of special equipment is treated always as if he were some sort of curious performing animal. The gunnery specialist, for example, may move and let off guns, but he may not say where they are to be let off–someone a little ignorant of range and trajectory does that; the engineer may move the ship and fire the battery, but only with some man, who does not perfectly understand, shouting instructions down a tube at him. If the cycle is to be adapted to military requirements, the thing is entrusted to Lieutenant-Colonel Balfour. If horses are to be bought for the British Army in India, no specialist goes, but Lord Edward Cecil. These people of the governing class do not understand there is such a thing as special knowledge or an inexorable fact in the world; they have been educated at schools conducted by amateur schoolmasters, whose real aim in life–if such people can be described as having a real aim in life–is the episcopal bench, and they have learnt little or nothing but the extraordinary power of appearances in these democratic times. To look right and to be of good report is to succeed. What else is there? The primarily functional men are ignored in the ostensible political scheme, it operates as though they did not exist, as though nothing, in fact, existed but the irresponsible wealthy, and the manipulators of irresponsible wealth, on the one hand, and a great, grey, politically indifferent community on the other. Having regard only to the present condition of political life, it would seem as though this state of affairs must continue indefinitely, and develop only in accordance with the laws of inter-action between our charlatan governing class on the one hand, and the grey mass of governed on the other. There is no way apparent in the existing political and social order, whereby the class of really educated persons that the continually more complicated mechanical fabric of social life is developing may be expected to come in. And in a very great amount of current political speculation, the development and final emergence of this class is ignored, and attention is concentrated entirely upon the inherent process of development of the political machine. And even in that it is very easy to exaggerate the preponderance of one or other of what are really very evenly balanced forces in the machine of democratic government.

There are two chief sets of parts in the machine that have a certain antagonistic relation, that play against each other, and one's conception of coming developments is necessarily determined by the relative value one gives to these opposing elements. One may compare these two groups to the Power and the Work, respectively, at the two ends of a lever.* On the one hand there is that which pays for the machine, which distributes salaries and rewards, subsidises newspapers and so forth–the central influence.** On the other hand, there is the collectively grey voting mass, with certain prejudices and traditions, and certain laws and limitations of thought upon which the newspapers work, and which, within the confines of its inherent laws, they direct. If one dwell chiefly on the possibilities of the former element, one may conjure up a practical end to democracy in the vision of a State "run" entirely by a group of highly forcible and intellectual persons–usually the dream takes the shape of financiers and their associates, their perfected mechanism of party control working the elections boldly and capably, and their public policy being directed towards financial ends. One of the common prophecies of the future of the United States is such a domination by a group of trust organisers and political bosses. But a man, or a group of men, so strong and intelligent as would be needed to hold an entire party machine within the confine of his–or their collective–mind and will, could, at the most, be but a very transitory and incidental phenomenon in the history of the world. Either such an exploitation of the central control will have to be covert and subtle beyond any precedent in human disingenuousness, or else its domination will have to be very amply modified indeed, by the requirements of the second factor, and its proceedings made very largely the resultant of that second factor's forces. Moreover, very subtle men do not aim at things of this sort, or aiming, fail, because subtlety of intelligence involves subtlety of character, a certain fastidiousness and a certain weakness. Now that the garrulous period, when a flow of language and a certain effectiveness of manner was a necessary condition to political pre-eminence, is passing away, political control falls

*The fulcrum, which is generally treated as being absolutely immovable, being the general belief in the theory of democracy.

**In the United States, a vast, rapidly developing country, with relatively much kinetic wealth, this central influence is the financial support of the Boss, consisting for the most part of active-minded, capable business organisers; in England, the land where irresponsible realised wealth is at a maximum, a public-spirited section of the irresponsible, inspired by the tradition of an aristocratic functional past, qualifies the financial influence with an amateurish, indolent, and publicly unprofitable integrity. In Germany an aggressively functional Court occupies the place and plays the part of a permanently dominant party machine.

more and more entirely into the hands of a barristerish, intriguing sort of person with a tough-wearing, leathery, practical mind. The sort of people who will work the machine are people with "faith," as the popular preachers say, meaning, in fact, people who do not analyse, people who will take the machine as it is, unquestioningly, shape their ambitions to it, and–saving their vanity–work it as it wants to go. The man who will be boss will be the man who wants to be boss, who finds, in being boss, a complete and final satisfaction, and not the man who complicates things by wanting to be boss in order to be, or do, something else. The machines are governed to-day, and there is every reason to believe that they will continue to be governed, by masterful-looking resultants, masters of nothing but compromise, and that little fancy of an inner conspiracy of control within the machine and behind ostensible politics is really on all fours with the wonderful Rodin (of the Juif Errant) and as probable as anything else in the romances of Eugene Sue.

If, on the other hand, we direct attention to the antagonistic element in the machine, to Public Opinion, to the alleged collective mind of the grey mass, and consider how it is brought to believe in itself and its possession of certain opinions by the concrete evidence of daily newspapers and eloquent persons saying as much, we may also very readily conjure up a contrasted vision of extraordinary demagogues or newspaper syndicates working the political machine from that direction. So far as the demagogue goes, the increase of population, the multiplication of amusements and interests, the differentiation of social habits, the diffusion of great towns, all militate against the sufficient gathering of masses of voters in meeting-houses which gave him his power in the recent past. It is improbable that ever again will any flushed undignified man with a vast voice, a muscular face in incessant operation, collar crumpled, hair disordered, and arms in wild activity, talking, talking, talking, talking copiously out of the windows of railway-carriages, talking on railway-platforms, talking from hotel balconies, talking on tubs, barrels, scaffoldings, pulpits–tireless and undammable–rise to be the most powerful thing in any democratic state in the world. Continually the individual vocal demagogue dwindles, and the element of bands and buttons, the organisation of the press and procession, the share of the machine, grows.

Mr. Harmsworth, of the London *Daily Mail,* in a very interesting article has glanced at certain possibilities of power that may vest in the owners of a great system of world-wide "simultaneous" newspapers, but he does not analyse the nature of the influence exercised by newspapers during the successive phases of the nineteenth century, nor the probable modifications of that influence in the years to come, and I think, on the whole, he inclines very naturally to overestimate the amount of intentional direction that may be given by the owner of a paper to the minds and acts of his readers, and to exceed the very definite limits within which that influence is confined. In the early Victorian period, the more limited, partly educated, and still very homogeneous enfranchised class, had a certain habit of thinking; its tranquil assurance upon most theological and all moral and æsthetic points left political questions as the chief field of exercise for such thinking as it did, and, as a consequence, the dignified newspapers of that time were able to discuss, and indeed were required to discuss not only specific situations but general principles. That indeed was their principal function, and it fell rather to the eloquent men to misapply these principles according to the necessity of the occasion. The papers did then very much more than they do now to mould opinion, though they did not direct affairs to anything like the extent of their modern successors. They made roads upon which events presently travelled in unexpected fashions. But the often cheaper and always more vivid newspapers that have come with the New Democracy do nothing to mould opinion. Indeed, there is no longer upon most public questions–and as I have tried to make clear in my previous paper, there is not likely to be any longer–a collective opinion to be moulded. Protectionists, for example, are a mere band, Free Traders are a mere band; on all these details we are in chaos. And these modern newspapers simply endeavour to sustain a large circulation and so merit advertisements by being as miscellaneously and vividly interesting as possible, by firing where the crowd seems thickest, by seeking perpetually and without any attempt at consistency, the greatest excitement of the greatest number. It is upon the cultivation and rapid succession of inflammatory topics that the modern newspaper expends its capital and trusts to recover its reward. Its general news sinks steadily to a subordinate position; criticism, discussion, and high responsibility pass out of journalism, and the power of the press comes more and more to be a dramatic and emotional power, the power to cry "Fire!" in the theatre, the power to give enormous value for a limited time to some personality, some event, some aspect, true or false, without any power of giving a specific direction to the forces this distortion may set going. Directly the press of to-day passes from that sort of thing to some specific proposal, some implication of principles and beliefs, directly it chooses and selects, then it passes from the miscellaneous to the sectarian, and out of touch with the grey indefiniteness of the general mind. It gives offence here, it perplexes and bores there; no more than the boss politician can the paper of great circulation afford to work consistently for any ulterior aim.

This is the limit of the power of the modern newspaper of large circulation, the newspaper that appeals to the grey element, to the average democratic man, the newspaper of the deliquescence, and if our previous conclusion that human society has ceased to be homogeneous and will presently display new masses segregating from a great confusion, holds good, that will be the limit of its power in the future. It may undergo many remarkable developments and modifications,* but none of these

*The nature of these modifications is an interesting side-issue. There is every possibility of papers becoming at last papers of world-wide circulation, so far as the language in which they are printed permits, with editions that will follow the sun and change into to-morrow's issue as they go, picking up literary criticism here, financial intelligence there, here to-morrow's story, and there to-morrow's scandal, and like some vast intellectual garden-roller, rolling out local provincialism at every revolution. This, for papers in English, at anyrate, is merely a question of how long it will be before the price of the best writing (for journalistic purposes) rises actually or relatively above the falling cost of long distance electrical typesetting. Each of the local editions of these world-travelling papers, in addition to the identical matter that will appear almost simultaneously everywhere, will no doubt have its special matter and its special advertisements. Illustrations will be telegraphed just as well as matter, and probably a much greater use will be made of sketch and diagram than at present. If the theory advanced in this book that democracy is a transitory confusion be sound, there will not be one world-paper of this sort only–like Moses' serpent after its miraculous struggle–but several, and as the non-provincial segregation of society goes on, these various great papers will take on more and more decided specific characteristics, and lose more and more their local references. They will come to have not only a distinctive type of matter, a distinctive method of thought and manner of expression, but distinctive fundamental implications, and a distinctive class of writer. This difference in character and tone renders the advent of any Napoleonic master of the newspaper world vastly more improbable than it would otherwise be. These specialising newspapers will, as they find their class, throw out many features that do not belong to that class. It is highly probable that many will restrict the space devoted to news and sham news; that forged and inflated stuff made in offices, that bulks out the foreign intelligence of so many English papers, for example. At present every paper contains a little of everything, inadequate sporting stuff, inadequate financial stuff, vague literary matter, voluminous reports of political vapourings, because no newspaper is quite sure of the sort of readers it has–probably no daily newspaper has yet a distinctive sort of reader.

Many people, with their minds inspired by the number of editions which evening papers pretend to publish and do not, incline to believe that daily papers may presently give place to hourly papers, each with the last news of the last sixty minutes photographically displayed. As a matter of fact no human being wants that, and very few are so foolish as to think they do; the only kind of news that any sort of people clamours for hot and hot is financial and betting fluctuations, lottery lists and examination results; and the elaborated and cheapened telegraphic and telephonic system of the coming days, with tapes (or phonograph to replace them) in every post-office and nearly every private house, so far from expanding this department, will probably sweep it out of the papers altogether. One will subscribe to a news-agency which will wire all the stuff one cares to have so violently fresh, into a phonographic recorder perhaps, in some convenient corner. There the thing will be in every house, beside the barometer, to hear or ignore. With the separation of that function what is left of the newspaper will revert to one daily edition–daily, I think, because of the power of habit to make the newspaper the specific business of some definite moments in the day; the breakfast-hour, I suppose, or the "up-to-town" journey with most Englishmen now. Quite possibly someone will discover some day that there is now machinery for folding and fastening a paper into a form that will not inevitably get into the butter, or lead to bitterness in a railway-carriage. This pitch of development reached, I incline to anticipate daily papers much more like the *Spectator* in form than these present mainsails of our public life. They will probably not contain fiction at all, and poetry only rarely, because no one but a partial imbecile wants these things in punctual daily doses, and we are anticipating an escape from a period of partial imbecility. My own culture and turn of mind, which is probably akin to that of a respectable mechanic of the year 2000, inclines me towards a daily paper that will have in addition to its concentrated and absolutely trustworthy daily news, full and luminous accounts of new inventions, new theories, and new departures of all sorts (usually illustrated), witty and penetrating comments upon public affairs, criticisms of all sorts of things, representations of newly produced works of art, and an ample amount of ably written controversy upon everything under the sun. The correspondence columns, instead of being an exercising place for bores and conspicuous people who are not mercenary, will be the most ample, the most carefully collected, and the most highly paid of all departments in this paper. Personal paragraphs will be relegated to some obscure and costly corner next to the births, deaths, and marriages. This paper will have, of course, many pages of business advertisements, and these will usually be well worth looking through, for the more intelligent editors of the days to come will edit this department just like any other, and classify their advertisements in a descending scale of freshness and interest that will also be an ascending scale of price. The advertiser who wants to be an indecent bore, and vociferate for the ten millionth time some flatulent falsehood about a pill, for instance, will pay at nuisance rates. Probably many papers will refuse to print nasty and distressful advertisements about people's insides at all. The entire paper will be as free from either greyness or offensive stupidity in its advertisement columns as the shop-windows in Bond Street to-day, and for much the same reason,–because the people who go that way do not want that sort of thing.

It has been supposed that, since the real income of the newspaper is derived from advertisements, large advertisers will combine in the future to own papers confined to the advertisements of their specific wares. Some such monopoly is already attempted; several publishing firms own or partially own a number of provincial papers, which they adorn with strange "Book Chat" tend to give it any greater political importance than it has now. And so, after all, our considerations of the probable developments of the party machine give us only negative results, so long as the grey social confusion continues. Subject to that continuance the party machine will probably continue as it is at present, and Democratic States and governments follow the lines upon which they run at the present time.

Now, how will the emergent class of capable men presently begin to modify the existing form of government in the ostensibly democratic countries and democratic monarchies? There will be very many variations and modifications of the methods of this arrival, an infinite complication of detailed incidents, but a general proposition will be found to hold good. The suppression of the party machine in the purely democratic countries and of the official choice of the rich and privileged rulers in the more monarchical ones, by capable operative and administrative men inspired by the belief in a common theory of social order, will come about–peacefully and gradually as a process of change, or violently as a revolution–but inevitably as the outcome of either of the imminence or else of the disasters of war.

columns conspicuously deficient in their information; and a well-known cycle-tyre firm supplies "Cycling" columns that are mere pedestals for the Head-of-King-Charles make of tyre. Many quack firms publish and give away annual almanacks replete with economical illustrations, offensive details, and bad jokes. But I venture to think, in spite of such phenomena, that these suggestions and attempts are made with a certain disregard of the essential conditions of sound advertisement. Sound advertisement consists in perpetual alertness and newness, in appearance in new places and in new aspects, in the constant access to fresh minds. The devotion of a newspaper to the interest of one particular make of a commodity or group of commodities will inevitably rob its advertisement department of most of its interest for the habitual readers of the paper. That is to say, the newspaper will fail in what is one of the chief attractions of a good newspaper. Moreover, such a devotion will react upon all the other matter in the paper, because the editor will need to be constantly alert to exclude seditious reflections upon the Health-Extract-of-Horse-Flesh or Saved-by-Boiling-Jam. His sense of this relation will taint his self-respect and make him a less capable editor than a man whose sole affair is to keep his paper interesting. To these more interesting rival papers the excluded competitor will be driven, and the reader will follow in his wake. There is little more wisdom in the proprietor of an article in popular demand buying or creating a newspaper to contain all his advertisements than in his buying a coal pit for the same purpose. Such a privacy of advertisement will never work, I think, on a large scale; it is probably at or near its maximum development now, and this anticipation of the advertiser-owned paper, like that of hourly papers, and that wonderfully powerful cosmic newspaper syndicate, is simply another instance of prophesying based only on a present trend, an expansion of the obvious, instead of an analysis of determining forces.

That all these governments of confusion will drift towards war, with a spacious impulse and a final vehemence quite out of comparison greater than the warlike impulses of former times, is a remarkable but no mean inexplicable thing. A tone of public expression, jealous and patriotic to the danger-point, is an unavoidable condition under which democratic governments exist. To be patriotically quarrelsome is imperative upon the party machines that will come to dominate the democratic countries. They will not possess detailed and definite policies and creeds because there are no longer any detailed and definite public opinions, but they will for all that require some ostensible purpose to explain their cohesion, some hold upon the common man that will ensure his appearance in numbers at the polling place sufficient to save the government from the raids of small but determined sects. That hold can be only of one sort. Without moral or religious uniformity, with material interests as involved and confused as a heap of spelicans, there remains only one generality for the politician's purpose, the ampler aspect of a man's egotism, his pride in what he imagines to be his particular kind–his patriotism. In every country amenable to democratic influences there emerges, or will emerge, a party machine, vividly and simply patriotic–and indefinite upon the score of any other possible consideration between man and man. This will hold true, not only of the ostensibly democratic states, but also of such reconstituted modern monarchies as Italy and Germany, for they, too, for all their legal difference, rest also on the grey. The party conflicts of the future will turn very largely on the discovery of the true patriot, on the suspicion that the crown or the machine in possession is in some more or less occult way traitorous, and almost all other matters of contention will be shelved and allowed to stagnate, for fear of breaking the unity of the national mechanism.

Now, patriotism is not a thing that flourishes in the void,–one needs a foreigner. A national and patriotic party is an anti-foreign party; the altar of the modern god, Democracy, will cry aloud for the stranger men. Simply to keep in power, and out of no love of mischief, the government or the party machine will have to insist upon dangers and national differences, to keep the voter to the poll by alarms, seeking ever to taint the possible nucleus of any competing organisation with the scandal of external influence. The party press will play the watch-dog and allay all internal dissensions with its warning bay at some adjacent people, and the adjacent peoples, for reasons to be presently expanded, will be continually more sensitive to such baying. Already one sees country yelping at country all over the modern world, not only in the matter of warlike issues, but with a note of quite furious commercial rivalry–quite furious and, indeed, quite insane, since its ideal of trading enormously with absolutely ruined and tradeless foreigners, exporting everything and importing nothing, is obviously outside reason altogether. The inexorable doom of these governments based on the grey, is to foster enmity between people and people. Even their alliances are but sacrifices to intenser antagonisms. And the phases of the democratic sequence are simple and sure. Forced on by a relentless competition, the tone of the outcries will become fiercer and fiercer; the occasions of excitement, the perilous moments, the ingenuities of annoyance, more and more dramatic,–from the mere emptiness and disorder of the general mind! Jealousies and anti-foreign enactments, tariff manipulations and commercial embitterment, destructive, foolish, exasperating obstructions that benefit no human being, will minister to this craving without completely allaying it. Nearer, and ever nearer, the politicians of the coming times will force one another towards the verge, not because they want to go over it, not because anyone wants to go over it, but because they are, by their very nature, compelled to go that way, because to go in any other direction is to break up and lose power. And, consequently, the final development of the democratic system, so far as intrinsic forces go, will be, not the rule of the boss, nor the rule of the trust, nor the rule of the newspaper; no rule, indeed, but international rivalry, international competition, international exasperation and hostility, and at last–irresistible and overwhelming–the definite establishment of the rule of that most stern and educational of all masters–*War.*

At this point there opens a tempting path, and along it historical precedents, like a forest of notice-boards, urge us to go. At the end of the vista poses the figure of Napoleon with "Cæsarism" written beneath it. Disregarding certain alien considerations for a time, assuming the free working out of democracy to its conclusion, we perceive that, in the case of our generalised state, the party machine, together with the nation entrusted to it, must necessarily be forced into passionate national war. But, having blundered into war, the party machine will have an air of having accomplished its destiny. A party machine or a popular government is surely as likely a thing to cause a big disorder of war and as unlikely a thing to conduct it, as the wit of man, working solely to that end, could ever have devised. I have already pointed out why we can never expect an elected government of the modern sort to be guided by any far-reaching designs, it is constructed to get office and keep office, not to do anything in office, the conditions of its survival are to keep appearances up and taxes down,* and the care and management of army and navy is quite outside its possibilities. The military and naval professions in our typical modern State will subsist very largely upon tradition, the ostensible government will interfere with rather than direct them, and there will be no force in the entire scheme to check the corrupting influence of a long peace, to insist upon adequate exercises for the fighting organisation or ensure an adequate adaptation to the new and perpetually changing possibilities of untried apparatus. Incapable but confident and energetic persons, having political influence, will have been permitted to tamper with the various arms of the service, the equipment will be largely devised to create an impression of efficiency in times of peace in the minds of the general voting public, and the really efficient soldiers will either have fretted themselves out of the army or have been driven out as political non-effectives, troublesome, innovating persons anxious to spend money upon "fads." So armed, the New Democracy will blunder into war, and the opening stage of the next great war will be the catastrophic breakdown of the formal armies, shame and disasters, and a disorder of conflict between more or less equally matched masses of stupefied, scared, and

*One striking illustration of the distinctive possibilities of democratic government came to light during the last term of office of the present patriotic British Government. As a demonstration of patriotism large sums of money were voted annually for the purpose of building warships, and the patriotic common man paid the taxes gladly with a dream of irresistible naval predominance to sweeten the payment. But the money was not spent on warships; only a portion of it was spent, and the rest remained to make a surplus and warm the heart of the common man in his tax-paying capacity. This artful dodge was repeated for several years; the artful dodger is now a peer, no doubt abjectly respected, and nobody in the most patriotic party so far evolved is a bit the worse for it. In the organising expedients of all popular governments, as in the prospectuses of unsound companies, the disposition is to exaggerate the nominal capital at the expense of the working efficiency. Democratic armies and navies are always short, and probably will always be short, of ammunition, paint, training and reserve stores; battalions and ships, since they count as units, are over-numerous and go short-handed, and democratic army reform almost invariably works out to some device for multiplying units by fission, and counting men three times instead of twice in some ingenious and plausible way. And this must be so, because the sort of men who come inevitably to power under democratic conditions are men trained by all the conditions of their lives to so set appearances before realities as at last to become utterly incapable of realities.

infuriated people. Just how far the thing may rise from the value of an alarming and edifying incident to a universal catastrophe, depends upon the special nature of the conflict, but it does not alter the fact that any considerable war is bound to be a bitter, appalling, highly educational and constitution-shaking experience for the modern democratic state.

Now, foreseeing this possibility, it is easy to step into the trap of the Napoleonic precedent. One hastens to foretell that either with the pressure of coming war, or in the hour of defeat, there will arise the Man. He will be strong in action, epigrammatic in manner, personally handsome and continually victorious. He will sweep aside parliaments and demagogues, carry the nation to glory, reconstruct it as an empire, and hold it together by circulating his profile and organising further successes. He will–I gather this from chance lights upon contemporary anticipations–codify everything, rejuvenate the papacy, or, at anyrate, galvanise Christianity, organise learning in meek, intriguing academies of little men, and prescribe a wonderful educational system. The grateful nations will once more deify a lucky and aggressive egotism. . . . And there the vision loses breath.

Nothing of the sort is going to happen, or, at anyrate, if it happens, it will happen as an interlude, as no necessary part in the general progress of the human drama. The world is no more to be recast by chance individuals than a city is to be lit by skyrockets. The purpose of things emerges upon spacious issues, and the day of individual leaders is past. The analogies and precedents that lead one to forecast the coming of military one-man-dominions, the coming of such other parodies of Cæsar's career as that misapplied, and speedily futile chess champion, Napoleon I. contrived, are false. They are false because they ignore two correlated things; first, the steady development of a new and quite unprecedented educated class as a necessary aspect of the expansion of science and mechanism, and secondly, the absolute revolution in the art of war that science and mechanism are bringing about. This latter consideration the next chapter will expand, but here, in the interests of this discussion, we may in general terms anticipate its gist. War in the past has been a thing entirely different in its nature from what war, with the apparatus of the future, will be–it has been showy, dramatic, emotional, and restricted; war in the future will be none of these things. War in the past was a thing of days and heroism; battles and campaigns rested in the hand of the great commander, he stood out against the sky, picturesquely on horseback, visibly controlling it all. War in the future will be a question of preparation, of long years of foresight and disciplined imagination, there will be no decisive victory, but a vast diffusion of conflict–it will depend less and less on controlling personalities and driving emotions, and more and more upon the intelligence and personal quality of a great number of skilled men. All this the next chapter will expand. And either before or after, but, at anyrate, in the shadow of war, it will become apparent, perhaps even suddenly, that the whole apparatus of power in the country is in the hands of a new class of intelligent and scientifically-educated men. They will probably, under the development of warlike stresses, be discovered–they will discover themselves–almost surprisingly with roads and railways, carts and cities, drains, food supply, electrical supply, and water supply, and with guns and such implements of destruction and intimidation as men scarcely dream of yet, gathered in their hands. And they will be discovered, too, with a growing common consciousness of themselves as distinguished from the grey confusion, a common purpose and implication that the fearless analysis of science is already bringing to light. They will find themselves with bloodshed and horrible disasters ahead, and the material apparatus of control entirely within their power. "Suppose, after all," they will say, "we ignore these very eloquent and showy governing persons above, and this very confused and ineffectual multitude below. Suppose now we put on the brakes and try something a little more stable and orderly. These people in possession have, of course, all sorts of established rights and prescriptions; they have squared the law to their purpose, and the constitution does not know us; they can get at the judges, they can get at the newspapers, they can do all sorts of things except avoid a smash–but, for our part, we have these really most ingenious and subtle guns. Suppose instead of our turning them and our valuable selves in a fool's quarrel against the ingenious and subtle guns of other men akin to ourselves, we use them in the cause of the higher sanity, and clear that jabbering war tumult out of the streets." . . . There may be no dramatic moment for the expression of this idea, no moment when the new Cromwellism and the new Ironsides will come visibly face to face with talk and baubles, flags and patriotic dinner-bells; but, with or without dramatic moments, the idea will be expressed and acted upon. It will be made quite evident then, what is now indeed only a pious opinion, namely, that wealth is, after all, no ultimate Power at all but only an influence among aimless, police-guarded men.

So long as there is peace the class of capable men may be mitigated and gagged and controlled, and the ostensible present order may flourish still in the hands of that other class of men which deals with the appearances of things. But as some supersaturated solution will crystallise out with the mere shaking of its beaker, so must the new order of men come into visibly organised existence through the concussions of war. The charlatans can escape everything except war, but to the cant and violence of nationality, to the sustaining force of international hostility, they are ruthlessly compelled to cling, and what is now their chief support must become at last their destruction. And so it is I infer that, whether violently as a revolution or quietly and slowly, this grey confusion that is Democracy must pass away inevitably by its own inherent conditions, as the twilight passes, as the embryonic confusion of the cocoon creature passes, into the higher stage, into the higher organism, the world-state of the coming years.

VI. WAR IN THE TWENTIETH CENTURY

IN shaping anticipations of the future of war there arises a certain difficulty about the point of departure. One may either begin upon such broad issues as the preceding forecasts have opened, and having determined now something of the nature of the coming State and the force of his warlike inclination, proceed to speculate how this vast, ill-organised fourfold organism will fight; or one may set all that matter aside for a space, and having regard chiefly to the continually more potent appliances physical science offers the soldier, we may try to develop a general impression of theoretically thorough war, go from that to the nature of the State most likely to be superlatively efficient in such warfare, and so arrive at the conditions of survival under which these present governments of confusion will struggle one against the other. The latter course will be taken here. We will deal first of all with war conducted for its own sake, with a model army, as efficient as an imaginative training can make it, and with a model organisation for warfare of the State behind it, and then the experience of the confused modern social organism as it is impelled, in an uncongenial metamorphosis, towards this imperative and finally unavoidable efficient state, will come most easily within the scope of one's imagination.

The great change that is working itself out in warfare is the same change that is working itself out in the substance of the social fabric. The essential change in the social fabric, as we have analysed it, is the progressive supersession of the old broad labour base by elaborately organised mechanism, and the obsolescence of the once valid and necessary distinction of gentle and simple. In warfare, as I have already indicated, this takes the form of the progressive supersession of the horse and the private soldier–which were the living and sole engines of the old time–by machines, and the obliteration of old distinction between leaders, who pranced in a conspicuously dangerous and encouraging way into the picturesque incidents of battle, and the led, who cheered and charged and filled the ditches and were slaughtered in a wholesale dramatic manner. The old war was a matter of long, dreary marches, great hardships of campaigning, but also of heroic conclusive moments. Long periods of campings–almost always with an outbreak of pestilence–of marchings and retreats, much crude business of feeding and forage, culminated at last, with an effect of infinite relief, in an hour or so of "battle." The battle was always a very intimate, tumultuous affair, the men were flung at one another in vast, excited masses, in living fighting machines as it were, spears or bayonets flashed, one side or the other cased to prolong the climax, and the thing was over. The beaten force crumpled as a whole, and the victors as a whole pressed upon it. Cavalry with slashing sabres marked the crowning point of victory. In the later stages of the old warfare musketry-volleys were added to the physical impact of the contending regiments, and at last cannon, as a quite accessory method of breaking these masses of men. So you "gave battle" to and defeated our enemy's forces wherever encountered, and when you reached your objective in his capital the war was done. . . . The new war will probably have none of these features of the old system of fighting.

The revolution that is in progress from the old war to a new war, different in its entire nature from the old, is marked primarily by the steady progress in range and efficiency of the rifle and of the field-gun–and more particularly of the rifle. The rifle develops persistently from a clumsy implement, that any clown may learn to use in half a day, towards a very intricate mechanism, easily put out of order and easily misused, but of the most extraordinary possibilities in the hands of men of courage, character, and high intelligence. Its precision at long range has made the business of its care, loading and aim subsidiary to the far more intricate matter of its use in relation to the contour of the ground within its reach. Even its elaboration as an instrument is probably still incomplete. One can conceive it provided in the future with cross-thread telescopic sights, the focussing of which, corrected by some ingenious use of hygroscopic material, might even find the range, and so enable it to be used with assurance up to a mile or more. It will probably also take on some of the characters of the machine-gun. It will be used either for single shots or to quiver and send a spray of almost simultaneous bullets out of a magazine evenly and certainly, over any small area the rifleman thinks advisable. It will probably be portable by one man, but there is no reason really, except the bayonet tradition, the demands of which may be met in other ways, why it should be the instrument of one sole man. It will, just as probably, be slung with its ammunition and equipment upon bicycle wheels, and be the common care of two or more associated soldiers. Equipped with such a weapon, a single couple of marksmen even, by reason of smokeless powder and carefully chosen cover, might make themselves practically invisible, and capable of surprising, stopping, and destroying a visible enemy in quite considerable numbers who blundered within a mile of them. And a series of such groups of marksmen so arranged as to cover the arrival of reliefs, provisions, and fresh ammunition from the rear, might hold out against any visible attack for an indefinite period, unless the ground they occupied was searched very ably and subtly by some sort of gun having a range in excess of their rifle-fire. If the ground they occupied were to be properly tunnelled and trenched, even that might not avail, and there would be nothing for it but to attack them by an advance under cover either of the night or of darkness caused by smoke-shells, or by the burning of cover about their position. Even then they might be deadly with magazine fire at close quarters. Save for their liability to such attacks, a few hundreds of such men could hold positions of a quite vast extent, and a few thousands might hold a frontier. Assuredly a mere handful of such men could stop the most multitudinous attack or cover the most disorderly retreat in the world, and even when some ingenious, daring, and lucky night assault had at last ejected them from a position, dawn would simply restore to them the prospect of reconstituting in new positions their enormous advantage of defence.

The only really effective and final defeat such an attenuated force of marksmen could sustain, would be from the slow and circumspect advance upon it of a similar force of superior marksmen, creeping forward under cover of night or of smoke-shells and fire, digging pits during the snatches of cessation obtained in this way, and so coming nearer and nearer and getting a completer and completer mastery of the defender's ground until the approach of the defender's reliefs, food, and fresh ammunition ceased to be possible. Thereupon there would be nothing for it but either surrender or a bolt in the night to positions in the rear, a bolt that might be hotly followed if it were deferred too late.

Probably between contiguous nations that have mastered the art of war, instead of the pouring clouds of cavalry of the old dispensation,* this will be the opening phase of the struggle, a vast duel all along the frontier between groups of skilled marksmen, continually being relieved and refreshed from the rear. For a time quite possibly there will be no definite army here or there, there will be no controllable battle, there will be no Great General in the field at all. But somewhere far in the rear the central organiser will sit at the telephonic centre of his fast front, and he will strengthen here and feed there and watch, watch perpetually the pressure, the incessant, remorseless pressure that is seeking to wear down his countervailing thrust. Behind the thin firing-line that is actually engaged, the country for many miles will be rapidly cleared and devoted to the business of war, big machines will be at work making second, third, and fourth lines of trenches that may be needed if presently the fire-line is forced back, spreading out transverse paths for the swift lateral movement of the cyclists who will be in perpetual alertness to relieve sudden local pressures, and all along those great motor roads our first "Anticipations" sketched, there will be a vast and rapid shifting to and fro of big and very long-range guns. These guns will

*Even along such vast frontiers as the Russian and Austrian, for example, where M. Bloch anticipates war will be begun with an invasion of clouds of Russian cavalry and great cavalry battles, I am inclined to think this deadlock of essentially defensive marksmen may still be the more probable thing. Small bodies of cyclist riflemen would rush forward to meet the advancing clouds of cavalry, would drop into invisible ambushes, and announce their presence–in unknown numbers–with carefully aimed shots difficult to locate. A small number of such men would always begin their fight with a surprise at the most advantageous moment, and they would be able to make themselves very deadly against a comparatively powerful frontal attack. If at last the attack were driven home before supports came up to the defenders, they would still be able to cycle away, comparatively immune. To attempt even very wide flanking movements against such a snatched position would be simply to run risks of blundering upon similar ambushes. The clouds of cavalry would have to spread into thin lines at last and go forward with the rifle. Invading clouds of cyclists would be in no better case. A conflict of cyclists against cyclists over a country too spacious for unbroken lines, would still, I think, leave the struggle essentially unchanged. The advance of small unsupported bodies would be the wildest and most unprofitable adventure; every advance would have to be made behind a screen of scouts, and, given a practical equality in the numbers and manhood of the two forces, these screens would speedily become simply very attenuated lines.

probably be fought with the help of balloons. The latter will hang above the firing-line all along the front, incessantly ascending and withdrawn; they will be continually determining the distribution of the antagonist's forces, directing the fire of continually shifting great guns upon the apparatus and supports in the rear of his fighting-line, forecasting his night plans and seeking some tactical or strategic weakness in that sinewy line of battle.

It will be evident that such warfare as this inevitable precision of gun and rifle forces upon humanity, will become less and less dramatic as a whole, more and more as a whole a monstrous thrust and pressure of people against people. No dramatic little general spouting his troops in the proper hysterics for charging, no prancing merely brave officers, no reckless gallantry or invincible stubbornness of men will suffice. For the commander-in-chief on a picturesque horse sentimentally watching his "boys" march past to death or glory in battalions, there will have to be a loyal staff of men, working simply, earnestly, and subtly to keep the front tight, and at the front, every little isolated company of men will have to be a council of war, a little conspiracy under the able man its captain, as keen and individual as a football team, conspiring against the scarcely seen company of the foe over yonder. The battalion commander will be replaced in effect by the organiser of the balloons and guns by which his few hundreds of splendid individuals will be guided and reinforced. In the place of hundreds of thousands of more or less drunken and untrained young men marching into battle–muddle-headed, sentimental, dangerous and futile hobbledehoys–there will be thousands of sober men braced up to their highest possibilities, intensely doing their best; in the place of charging battalions, shattering impacts of squadrons and wide harvest-fields of death, there will be hundreds of little rifle battles fought up to the hilt, gallant dashes here, night surprises there, the sudden sinister faint gleam of nocturnal bayonets, brilliant guesses that will drop catastrophic shell and death over hills and forests suddenly into carelessly exposed masses of men. For eight miles on either side of the firing-lines–whose fire will probably never altogether die away while the war lasts–men will live and eat and sleep under the imminence of unanticipated death. . . . Such will be the opening phase of the war that is speedily to come.

And behind the thin firing-line on either side a vast multitude of people will be at work; indeed, the whole mass of the efficients in the State will have to be at work, and most of them will be simply at the same work or similar work to that done in peace time–only now as combatants upon the lines of communication. The organised staffs of the big road managements, now become a part of the military scheme, will be deporting women and children and feeble people and bringing up supplies and supports; the doctors will be dropping from their civil duties into pre-appointed official places, directing the feeding and treatment of the shifting masses of people and guarding the valuable manhood of the fighting apparatus most sedulously from disease;* the engineers will be entrenching and bringing up a vast variety of complicated and ingenious apparatus designed to surprise and inconvenience the enemy in novel ways; the dealers in food and clothing, the manufacturers of all sorts of necessary stuff, will be converted by the mere declaration of war into public servants; a practical realisation of socialistic conceptions will quite inevitably be forced upon the fighting State. The State that has not incorporated with its fighting organisation all its able-bodied manhood and all its material substance, its roads, vehicles, engines, foundries, and all its resources of food and clothing; the State which at the outbreak of war has to bargain with railway and shipping companies, replace experienced station-masters by inexperienced officers, and haggle against alien interests for every sort of supply, will be at an overwhelming disadvantage against a State which has emerged from the social confusion of the present time, got rid of every vestige of our present distinction between official and governed, and organised every element in its being.

I imagine that in this ideal war as compared with the war of to-day, there will be a very considerable restriction of the rights of the non-combatant. A large part of existing International Law involves a curious implication, a distinction between the belligerent government and its accredited agents in warfare and the general body of its subjects. There is a disposition to treat the belligerent government, in spite of the democratic status of many States, as not fully representing its people, to establish a sort of world-citizenship in the common mass outside the official and military class. Protection of the non-combatant and his property comes at last–in theory at least–within a measurable distance of notice-boards: "Combatants are requested to keep off the grass."

*So far, pestilence has been a feature of almost every sustained war in the world, but there is really no reason whatever why it should be so. There is no reason, indeed, why a soldier upon active service on the victorious side should go without a night's rest or miss a meal. If he does, there is muddle and want of foresight somewhere, and that our hypothesis excludes.

This disposition I ascribe to a recognition of that obsolescence and inadequacy of the formal organisation of States, which has already been discussed in this book. It was a disposition that was strongest perhaps in the earliest decades of the nineteenth century, and stronger now than, in the steady and irresistible course of strenuous and universal military preparation, it is likely to be in the future. In our imaginary twentieth century State, organised primarily for war, this tendency to differentiate a non-combatant mass in the fighting State will certainly not be respected, the State will be organised as a whole to fight as a whole, it will have triumphantly asserted the universal duty of its citizens. The military force will be a much ampler organisation than the "army" of to-day, it will be not simply the fists but the body and brain of the land. The whole apparatus, the whole staff engaged in internal communication, for example, may conceivably not be State property and a State service, but if it is not it will assuredly be as a whole organised as a volunteer force, that may instantly become a part of the machinery of defence or aggression at the outbreak of war.* The men may very conceivably not have a uniform, for military uniforms are simply one aspect of this curious and transitory phase of restriction, but they will have their orders and their universal plan. As the bells ring and the recording telephones click into every house the news that war has come, there will be no running to and fro upon the public ways no bawling upon the moving platforms of the central urban nuclei, no crowds of silly, useless able-bodied people gaping at inflammatory transparencies outside the offices of sensational papers because the egregious idiots in control of affairs have found them no better employment. Every man will be soberly and intelligently setting about the particular thing he has to do—even the rich shareholding sort of person, the hereditary mortgager of society, will be given something to do, and if he has learnt nothing else he will serve to tie up parcels of ammunition or pack army sausage. Very probably the best of such people and of the speculative class will have qualified as cyclist marksmen for the front, some of them may even have devoted the leisure of peace to military studies and may be prepared with

*Lady Maud Rolleston, in her very interesting *Yeoman Service,* complains of the Boers killing an engine-driver during an attack on a train at Kroonstadt, "which was," she writes, "an abominable action, as he is, in law, a non-combatant." The implicit assumption of this complaint would cover the engineers of an ironclad or the guides of a night attack, everybody, in fact, who was not positively weapon in hand.

novel weapons. Recruiting among the working classes–or, more properly speaking, among the People of the Abyss–will have dwindled to the vanishing-point; people who are no good for peace purposes are not likely to be any good in such a grave and complicated business as modern war. The spontaneous traffic of the roads in peace, will fall now into two streams, one of women and children coming quietly and comfortably out of danger, the other of men and material going up to the front. There will be no panics, no hardships, because everything will have been amply pre-arranged–we are dealing with an ideal State. Quietly and tremendously that State will have gripped its adversary and tightened its muscles–that is all.

Now the strategy of this new sort of war in its opening phase will consist mainly in very rapid movements of guns and men behind that thin screen of marksmen, in order to deal suddenly and unexpectedly some forcible blow, to snatch at some position into which guns and men may be thrust to outflank and turn the advantage of the ground against some portion of the enemy's line. The game will be largely to crowd and crumple that line, to stretch it over an arc to the breaking-point, to secure a position from which to shell and destroy its supports and provisions, and to capture or destroy its guns and apparatus, and so tear it away from some town or arsenal it has covered. And a factor of primary importance in this warfare, because of the importance of seeing the board, a factor which will be enormously stimulated to develop in the future, will be the aerial factor. Already we have seen the captive balloon as an incidental accessory of considerable importance even in the wild country warfare of South Africa. In the warfare that will go on in the highly-organised European States of the opening century, the special military balloon used in conjunction with guns, conceivably of small calibre but of enormous length and range, will play a part of quite primary importance. These guns will be carried on vast mechanical carriages, possibly with wheels of such a size as will enable them to traverse almost all sorts of ground.* The aeronauts, provided

*Experiments will probably be made in the direction of armoured guns, armoured search-light carriages, and armoured shelters for men, that will admit of being pushed forward over rifle-swept ground. To such possibilities, to possibilities even of a sort of land ironclad, my inductive reason inclines; the armoured train seems indeed a distinct beginning of this sort of thing, but my imagination proffers nothing but a vision of wheels smashed by shells, iron tortoises gallantly rushed by hidden men, and unhappy marksmen and engineers being shot at as they bolt from some such monster overset. The fact of it is, I detest and fear these thick, slow, essentially defensive methods, either for land

with large scale-maps of the hostile country, will mark down to the gunners below the precise point upon which to direct their fire, and over hill and dale the shell will fly–ten miles it may be–to its billet, camp, massing night attack, or advancing gun.

Great multitudes of balloons will be the Argus eyes of the entire military organism, stalked eyes with a telephonic nerve in each stalk, and at night they will sweep the country with search-lights and come soaring before the wind with hanging flares. Certainly they will be steerable. Moreover, when the wind admits, there will be freely-moving steerable balloons wagging little flags to their friends below. And so far as the resources of the men on the ground go, the balloons will be almost invulnerable. The mere perforation of balloons with shot does them little harm, and the possibility of hitting a balloon that is drifting about at a practically unascertainable distance and height so precisely as to blow it to pieces with a timed shell, and to do this in the little time before it is able to give simple and precise instructions as to your range and position to the unseen gunners it directs, is certainly one of the most difficult and trying undertakings for an artilleryman that one can well imagine. I am inclined to think that the many considerations against a successful attack on balloons from the ground, will enormously stimulate enterprise and invention in the direction of dirigible aerial devices that can fight. Few people, I fancy, who know the work of Langley, Lilienthal, Pilcher, Maxim, and Chanute, but will be inclined to believe that long before the year A.D. 2000, and very probably before 1950, a successful aeroplane will have soared and come home safe and sound. Directly that is accomplished the new invention will be most assuredly applied to war.

The nature of the things that will ultimately fight in the sky is a matter for curious speculation. We begin with the captive balloon. Against that the navigable balloon will presently operate. I am inclined to think the practicable navigable balloon will be first attained by the use of a device already employed by Nature in the swimming-bladder of fishes. This is a closed gas-bag that can be contracted or expanded. If a gas-bag of thin, strong, practically impervious substance could be enclosed in a net of closely interlaced fibres (interlaced, for example, on the pattern of the muscles of the bladder in mammals), the ends of these fibres might be wound and unwound, and the effect of contractility attained.

or sea fighting. I believe invincibly that the side that can go fastest and hit hardest will always win, with or without or in spite of massive defences, and no ingenuity in devising the massive defence will shake that belief.

A row of such contractile balloons, hung over a long car which was horizontally expanded into wings, would not only allow that car to rise and fall at will, but if the balloon at one end were contracted and that at the other end expanded, and the intermediate ones allowed to assume intermediate conditions, the former end would drop, the expanded wings would be brought into a slanting condition over a smaller area of supporting air, and the whole apparatus would tend to glide downwards in that direction. The projection of a small vertical plane upon either side would make the gliding mass rotate in a descending spiral, and so we have all the elements of a controllable flight. Such an affair would be difficult to overset. It would be able to beat up even in a fair wind, and then it would be able to contract its bladders and fall down a long slant in any direction. From some such crude beginning a form like a soaring, elongated, flat-brimmed hat might grow, and the possibilities of adding an engine-driven screw are obvious enough.

It is difficult to see how such a contrivance could carry guns of any calibre unless they fired from the rear in the line of flight. The problem of recoil becomes a very difficult one in aerial tactics. It would probably have at most a small machine-gun or so, which might fire an explosive shell at the balloons of the enemy, or kill their aeronauts with distributed bullets. The thing would be a sort of air-shark, and one may even venture to picture something of the struggle the deadlocked marksmen of 1950, lying warily in their rifle-pits, will see.

One conceives them at first, each little hole with its watchful, well-equipped couple of assassins, turning up their eyes in expectation. The wind is with our enemy, and his captive balloons have been disagreeably overhead all through the hot morning. His big guns have suddenly become nervously active. Then, a little murmur along the pits and trenches, and from somewhere over behind us, this air-shark drives up the sky. The enemy's balloons splutter a little, retract, and go rushing down, and we send a spray of bullets as they drop. Then against our aerostat, and with the wind driving them clean overhead of us, come the antagonistic flying-machines. I incline to imagine there will be a steel prow with a cutting edge at either end of the sort of aerostat I foresee, and conceivably this aerial ram will be the most important weapon of the affair. When operating against balloons, such a fighting-machine will rush up the air as swiftly as possible, and then, with a rapid contraction of its bladders, fling itself like a knife at the sinking war-balloon of the foe.

Down, down, down, through a vast alert tension of flight, down it will swoop, and, if its stoop is successful, slash explosively at last through a suffocating moment. Rifles will crack, ropes tear and snap; there will be a rending and shouting, a great thud of liberated gas, and perhaps a flare. Quite certainly those flying-machines will carry folded parachutes, and the last phase of many a struggle will be the desperate leap of the aeronauts with these in hand, to snatch one last chance of life out of a mass of crumpling, fallen wreckage.

But in such a fight between flying-machine and flying-machine as we are trying to picture, it will be a fight of hawks, complicated by bullets and little shells. They will rush up and up to get the pitch of one another, until the aeronauts sob and sicken in the rarefied air, and the blood comes to the eyes and nails. The marksmen below will strain at last, eyes under hands, to see the circling battle that dwindles in the zenith. Then, perhaps, a wild adventurous dropping of one close beneath the other, an attempt to stoop, the sudden splutter of guns, a tilting up or down, a disengagement. What will have happened? One combatant, perhaps, will heel lamely earthward, dropping, dropping, with half its bladders burst or shot away, the other circles down in pursuit. . . . "What are they doing?" Our marksmen will snatch at their field-glasses, tremulously anxious, "Is that a white flag or no? . . . If they drop now we have 'em!"

But the duel will be the rarer thing. In any affair of ramming there is an enormous advantage for the side that can contrive, anywhere in the field of action, to set two vessels at one. The mere ascent of one flying-ram from one side will assuredly slip the leashes of two on the other, until the manœuvring squadrons may be as thick as starlings in October. They will wheel and mount, they will spread and close, there will be elaborate manœuvres for the advantage of the wind, there will be sudden drops to the shelter of entrenched guns. The actual impact of battle will be an affair of moments. They will be awful moments, but not more terrible, not more exacting of manhood than the moments that will come to men when there is–and it has not as yet happened on this earth–equal fighting between properly manned and equipped ironclads at sea. (And the well-bred young gentlemen of means who are privileged to officer the British Army nowadays will be no more good at this sort of thing than they are at controversial theology or electrical engineering or anything else that demands a well-exercised brain.) . . .

Once the command of the air is obtained by one of the contending armies, the war must become a conflict between a seeing host and one that is blind. The victor in that aerial struggle will tower with pitilessly watchful eyes over his adversary, will concentrate his guns and all his strength unobserved, will mark all his adversary's roads and communications, and sweep them with sudden incredible disasters of shot and shell. The moral effect of this predominance will be enormous. All over the losing country, not simply at his frontier but everywhere, the victor will soar. Everybody everywhere will be perpetually and constantly looking up, with a sense of loss and insecurity, with a vague stress of painful anticipations. By day the victor's aeroplanes will sweep down upon the apparatus of all sorts in the adversary's rear, and will drop explosives and incendiary matters upon them,* so that no apparatus or camp or shelter will any longer be safe. At night his high, floating search-lights will go to and fro and discover and check every desperate attempt to relieve or feed the exhausted marksmen of the fighting-line. The phase of tension will pass, that weakening opposition will give, and the war from a state of mutual pressure and petty combat will develop into the collapse of the defensive lines. A general advance will occur under the aerial van, ironclad road fighting-machines may perhaps play a considerable part in this, and the enemy's line of marksmen will be driven back or starved into surrender, or broken up and hunted down. As the superiority of the attack becomes week by week more and more evident, its assaults will become more dashing and far-reaching. Under the moonlight and the watching balloons there will be swift, noiseless rushes of cycles, precipitate dismounts, and the never-to-be-quite-abandoned bayonet will play its part. And now men on the losing side will thank God for the reprieve of a pitiless wind, for lightning, thunder, and rain, for any elemental disorder that will for a moment lift the descending scale! Then, under banks of fog and cloud, the victorious advance will pause and grow peeringly watchful and nervous, and mud-stained desperate men will go splashing forward into an elemental blackness, rain or snow like a benediction on their faces, blessing the primordial savagery of nature that can still set aside the wisest devices of men, and give the unthrifty one last desperate chance to get their own again or die.

Such adventures may rescue pride and honour, may cause momentary dismay in the victor and palliate disaster, but they will not turn back the advance of the victors, or twist inferiority

*Or, in deference to the Rules of War, fire them out of guns of trivial carrying power.

into victory. Presently the advance will resume. With that advance the phase of indecisive contest will have ended, and the second phase of the new war, the business of forcing submission, will begin. This should be more easy in the future even than it has proved in the past, in spite of the fact that central governments are now elusive, and small bodies of rifle-armed guerrillas far more formidable than ever before. It will probably be brought about in a civilised country by the seizure of the vital apparatus of the urban regions–the water supply, the generating stations of electricity (which will supply all the heat and warmth of the land), and the chief ways used in food distribution. Through these expedients, even while the formal war is still in progress, an irresistible pressure upon a local population will be possible, and it will be easy to subjugate or to create afresh local authorities, who will secure the invader from any danger of a guerrilla warfare upon his rear. Through that sort of an expedient an even very obdurate loser will be got down to submission, area by area. With the destruction of its military apparatus and the prospective loss of its water and food supply, however, the defeated civilised State will probably be willing to seek terms as a whole, and bring the war to a formal close.

In cases where, instead of contiguous frontiers, the combatants are separated by the sea, the aerial struggle will probably be preceded or accompanied by a struggle for the command of the sea. Of this warfare there have been many forecasts. In this, as in all the warfare of the coming time, imaginative foresight, a perpetual alternation of tactics, a perpetual production of unanticipated devices, will count enormously. Other things being equal, victory will rest with the force mentally most active. What type of ship may chance to be prevalent when the great naval war comes is hard guessing, but I incline to think that the naval architects of the ablest peoples will concentrate more and more upon speed and upon range and penetration, and, above all, upon precision of fire. I seem to see a light type of ironclad, armoured thickly only over its engines and magazines, murderously equipped, and with a ram–as alert and deadly as a striking snake. In the battles of the open she will have little to fear from the slow, fumbling treacheries of the submarine, she will take as little heed of the chance of a torpedo as a barefooted man in battle does of the chance of a fallen dagger in his path. Unless I know nothing of my own blood, the English and Americans will prefer to catch their enemies in ugly weather or at night, and then they will fight to ram. The struggle on the high seas between any two naval powers (except, perhaps, the English and American, who have both quite unparalleled opportunities for coaling) will not last more than a week or so. One or other force will be destroyed at sea, driven into its ports and blockaded there, or cut off from its supply of coal (or other force-generator), and hunted down to fight or surrender. An inferior fleet that tries to keep elusively at sea will always find a superior fleet between itself and coal, and will either have to fight at once or be shot into surrender as it lies helpless on the water. Some commerce-destroying enterprise on the part of the loser may go on, but I think the possibilities of that sort of thing are greatly exaggerated. The world grows smaller and smaller, the telegraph and telephone go everywhere, wireless telegraphy opens wider and wider possibilities to the imagination, and how the commerce-destroyer is to go on for long without being marked down, headed off, cut off from coal, and forced to fight or surrender, I do not see. The commerce-destroyer will have a very short run; it will have to be an exceptionally good and costly ship in the first place, it will be finally sunk or captured, and altogether I do not see how that sort of thing will pay when once the command of the sea is assured. A few weeks will carry the effective frontier of the stronger power up to the coast-line of the weaker, and permit of the secure resumption of the over-sea trade of the former. And then will open a second phase of naval warfare, in which the submarine may play a larger part.

I must confess that my imagination, in spite even of spurring, refuses to see any sort of submarine doing anything but suffocate its crew and founder at sea. It must involve physical inconvenience of the most demoralising sort simply to be in one for any length of time. A first-rate man who has been breathing carbonic acid and oil vapour under a pressure of four atmospheres becomes presently a second-rate man. Imagine yourself in a submarine that has ventured a few miles out of port, imagine that you have headache and nausea, and that some ship of the *Cobra* type is flashing itself and its search-lights about whenever you come up to the surface, and promptly tearing down on your descending bubbles with a ram, trailing perhaps a tail of grapples or a net as well. Even if you get their boat, these nicely aerated men you are fighting know they have a four to one chance of living; while for your submarine to be "got" is certain death. You may, of course, throw out a torpedo or so, with as much chance of hitting vitally as you would have if you were blindfolded, turned round three times, and told to fire revolver-shots at a charging elephant. The possibility of sweeping for a submarine with a seine would be vividly present in the minds of a submarine crew. If you are near shore you will probably be near rocks–an unpleasant complication in a hurried dive. There would, probably, very soon be boats out too, seeking with a machine-gun or pompom for a chance at your occasionally emergent conning-tower. In no way can a submarine be more than purblind, it will be, in fact, practically blind. Given a derelict ironclad on a still night within sight of land, a carefully handled submarine might succeed in groping its way to it and destroying it; but then it would be much better to attack such a vessel and capture it boldly with a few desperate men on a tug. At the utmost the submarine will be used in narrow waters, in rivers, or to fluster or destroy ships in harbour or with poor-spirited crews–that is to say, it will simply be an added power in the hands of the nation that is predominant at sea. And, even then, it can be merely destructive, while a sane and high-spirited fighter will always be dissatisfied if, with an indisputable superiority of force, he fails to take.*

No; the naval warfare of the future is for light, swift ships, almost recklessly not defensive and with splendid guns and gunners. They will hit hard and ram, and warfare which is taking to cover on land will abandon it at sea. And the captain, and the engineer, and the gunner will have to be all of the same sort of men: capable, headlong men, with brains and no ascertainable social position. They will differ from the officers of the British Navy in the fact that the whole male sex of the nation will have been ransacked to get them. The incredible stupidity that closes all but a menial position in the British Navy to the sons of those who cannot afford to pay a hundred a year for them for some years, necessarily brings the individual quality of the British naval officer below the highest possible, quite apart from the deficiencies that must exist on account of the badness of secondary education in England. The British naval officer and engineer are not made the best of, good as they are, indisputably they might be infinitely better both in quality and training. The smaller German navy, probably, has an ampler pick of men relatively, is far better educated, less confident, and more strenuous. But the abstract navy I am here writing of will be superior to either

*A curious result might very possibly follow a success of submarines on the part of a naval power finally found to be weaker and defeated. The victorious power might decide that a narrow sea was no longer, under the new conditions, a comfortable boundary line, and might insist on marking its boundary along the high-water mark of its adversary's adjacent coasts.

of these, and like the American, in the absence of any distinction between officers and engineers. The officer will be an engineer.

The military advantages of the command of the sea will probably be greater in the future than they have been in the past. A fleet with aerial supports would be able to descend upon any portion of the adversary's coast it chose, and to dominate the country inland for several miles with its gun-fire. All the enemy's sea-coast towns would be at its mercy. It would be able to effect landing and send raids of cyclist-marksmen inland, whenever a weak point was discovered. Landings will be enormously easier than they have ever been before. Once a wedge of marksmen has been driven inland they would have all the military advantages of the defence when it came to eject them. They might, for example, encircle and block some fortified post, and force costly and disastrous attempts to relieve it. The defensive country would stand at bay, tethered against any effective counter-blow, keeping guns, supplies, and men in perpetual and distressing movement to and fro along its sea-frontiers. Its soldiers would get uncertain rest, irregular feeding, unhealthy conditions of all sorts in hastily made camps. The attacking fleet would divide and re-unite, break up and vanish, amazingly reappear. The longer the defender's coast the more wretched his lot. Never before in the world's history was the command of the sea worth what it is now. But the command of the sea is, after all, like military predominance on land, to be insured only by superiority of equipment in the hands of a certain type of man, a type of man that it becomes more and more impossible to improvise, that a country must live for through many years, and that no country on earth at present can be said to be doing its best possible to make.

All this elaboration of warfare lengthens the scale between theoretical efficiency and absolute unpreparedness. There was a time when any tribe that had men and spears was ready for war, and any tribe that had some cunning or emotion at command might hope to discount any little disparity in numbers between itself and its neighbour. Luck and stubbornness and the incalculable counted for much; it was half the battle not to know you were beaten, and it is so still. Even to-day, a great nation, it seems, may still make its army the plaything of its gentlefolk, abandon important military appointments to feminine intrigue, and trust cheerfully to the homesickness and essential modesty of its influential people, and the simpler patriotism of its colonial dependencies when it comes at last to the bloody and wearisome business of "muddling through." But these days of the happy-go-lucky optimist are near their end. War is being drawn into the field of the exact sciences. Every additional weapon, every new complication of the art of war, intensifies the need of deliberate preparation, and darkens the outlook of a nation of amateurs. Warfare in the future, on sea or land alike, will be much more one-sided than it has ever been in the past, much more of a foregone conclusion. Save for national lunacy, it will be brought about by the side that will win, and because that side knows that it will win. More and more it will have the quality of surprise, of pitiless revelation. Instead of the see-saw, the bickering interchange of battles of the old time, will come swiftly and amazingly blow, and blow, and blow, no pause, no time for recovery, disasters cumulative and irreparable.

The fight will never be in practice between equal sides, never be that theoretical deadlock we have sketched, but a fight between the more efficient and the less efficient, between the more inventive and the more traditional. While the victors, disciplined and grimly intent, full of the sombre yet glorious delight of a grave thing well done, will, without shouting or confusion, be fighting like one great national body, the losers will be taking that pitiless exposure of helplessness in such a manner as their natural culture and character may determine. War for the losing side will be an unspeakably pitiable business. There will be first of all the coming of the war, the wave of excitement, the belligerent shouting of the unemployed inefficients, the flag-waving, the secret doubts, the eagerness for hopeful news, the impatience of the warning voice. I seem to see, almost as if he were symbolic, the grey old general–the general who learnt his art of war away in the vanished nineteenth century, the altogether too elderly general with his epaulettes and decorations, his uniform that has still its historical value, his spurs and his sword–riding along on his obsolete horse, by the side of his doomed column. Above all things he is a gentleman. And the column looks at him lovingly with its countless boys' faces, and the boys' eyes are infinitely trustful, for he has won battles in the old time. They will believe in him to the end. They have been brought up in their schools to believe in him and his class, their mothers have mingled respect for the gentlefolk with the simple doctrines of their faith, their first lesson on entering the army was the salute. The "smart" helmets His Majesty, or some such unqualified person, chose for them, lie hotly on their young brows, and over their shoulders slope their obsolete, carelessly-sighted guns. Tramp, tramp, they march doing what they have been told to do, incapable of doing anything they have not been told to do, trustful and pitiful, marching to wounds and disease, hunger, hardship, and death. They know nothing of what they are going to meet, nothing of what they will have to do; Religion and the Ratepayer and the Rights of the Parent working through the instrumentality of the Best Club in the World have kept their souls and minds, if not untainted, at least only harmlessly veneered, with the thinnest sham of training or knowledge. Tramp, tramp, they go, boys who will never be men, rejoicing patriotically in the nation that has thus sent them forth, badly armed, badly clothed, badly led, to be killed in some avoidable quarrel by men unseen. And beside them, an absolute stranger to them, a stranger even in habits of speech and thought, and at anyrate to be shot with them fairly and squarely, marches the subaltern–the son of the school-burking, shareholding class–a slightly taller sort of boy, as ill-taught as they are in all that concerns the realities of life, ignorant of how to get food, how to get water, how to keep fever down and strength up, ignorant of his practical equality with the men beside him, carefully trained under a clerical headmaster to use a crib, play cricket rather nicely, look all right whatever happens, believe in his gentility, and avoid talking "shop." . . . The major you see is a man of the world, and very pleasantly meets the grey general's eye. He is, one may remark by the way, something of an army reformer, without offence, of course, to the Court people or the Government people. His prospects–if only he were not going to be shot–are brilliant enough. He has written quite cleverly on the question of Recruiting, and advocated as much as twopence more a day and billiard-rooms under the chaplain's control; he has invented a military bicycle with a wheel of solid iron that can be used as a shield; and a war correspondent and, indeed, anyone who writes even the most casual and irresponsible article on military questions is a person worth his cultivating. He is the very life and soul of army reform, as it is known to the governments of the grey–that is to say, army reform without a single step towards a social revolution. . . .

So the gentlemanly old general–the polished drover to the shambles–rides, and his doomed column marches by, in this vision that haunts my mind.

I cannot foresee what such a force will even attempt to do, against modern weapons. Nothing can happen but the needless and most wasteful and pitiful killing of these poor lads, who make up the infantry battalions, the main mass of all the European armies of to-day, whenever they come against a sanely-organised army. There is nowhere they can come in, there is nothing they can do. The scattered, invisible marksmen with their supporting guns will shatter their masses, pick them off individually, cover their line of retreat and force them into wholesale surrenders. It will be more like herding sheep than actual fighting. Yet the bitterest and cruellest things will have to happen, thousands and thousands of poor boys will be smashed in all sorts of dreadful ways and given over to every conceivable form of avoidable hardship and painful disease, before the obvious fact that war is no longer a business for half-trained lads in uniform, led by parson-bred sixth-form boys and men of pleasure and old men, but an exhaustive demand upon very carefully-educated adults for the most strenuous best that is in them, will get its practical recognition.* . . .

*There comes to hand as I correct these proofs a very typical illustration of the atmosphere of really almost imbecile patronage in which the British private soldier lives. It is a circular from someone at Lydd, someone who evidently cannot even write English, but who is nevertheless begging for an iron hut in which to inflict lessons on our soldiers. "At present," says the circular, "it is pretty to see in the Home a group of Gunners busily occupied in wool-work or learning basket-making, whilst one of their number sings or recites, and others are playing games or letter-writing, but even quite recently the members of the Bible Reading Union and one of the ladies might have been seen painfully crowded behind screens, choosing the 'Golden Text' with lowered voices, and trying to pray 'without distraction,' whilst at the other end of the room men were having supper, and half-way down a dozen Irish militia (who don't care to read, but are keen on a story) were gathered round another lady, who was telling them an amusing temperance tale, trying to speak so that the Bible readers should not hear her and yet that the Leinsters *should* was a difficulty, but when the Irishmen begged for a song–difficulty became *impossibility,* and their friend had to say, *'No.'* Yet this is just the double work required in Soldiers' Homes, and above all at Lydd, where there is so little safe amusement to be had in camp, and none in the village." These poor youngsters go from this "safe amusement" under the loving care of "lady workers," this life of limitation, make-believe and spiritual servitude that a self-respecting negro would find intolerable, into a warfare that exacts initiative and a freely acting intelligence from all who take part in it, under the bitterest penalties of shame and death. What can you expect of them? And how can you expect any men of capacity and energy, any men even of mediocre self-respect to knowingly place themselves under the tutelage of the sort of people who dominate these organised degradations? I am amazed the army gets so many capable recruits as it does. And while the private lives under these conditions, the would-be capable officer stifles amidst equally impossible surroundings. He must associate with the uneducated products of the public schools, and listen to their chatter about the "sports" that delight them, suffer social indignities from the "army woman," worry and waste money on needless clothes, and expect to end by being shamed or killed under some unfairly promoted incapable. Nothing illustrates the intellectual blankness of the British army better than its absolute dearth of military literature. No one would dream of gaining any profit by writing or

Well, in the ampler prospect even this haunting tragedy of innumerable avoidable deaths is but an incidental thing. They die, and their troubles are over. The larger fact after all is the inexorable tendency in things to make a soldier a skilled and educated man, and to link him, in sympathy and organisation, with the engineer and the doctor, and all the continually developing mass of scientifically educated men that the advance of science and mechanism is producing. We are dealing with the inter-play of two world-wide forces, that work through distinctive and contrasted tendencies to a common end. We have the force of invention insistent upon a progress of the peace organisation, which tends on the one hand to throw out great useless masses of people, the People of the Abyss, and on the other hand to develop a sort of adiposity of functionless wealthy, a speculative elephantiasis, and to promote the development of a

publishing a book upon such a subject, for example, as mountain warfare in England, because not a dozen British officers would have the sense to buy such a book, and yet the British army is continually getting into scrapes in mountain districts. A few unselfish men like Major Peech find time to write an essay or so, and that is all. On the other hand, I find no less than five works in French on this subject in MM. Chapelet & Cie.'s list alone. On guerrilla warfare again, and after two years of South Africa, while there is nothing in English but some scattered papers by Dr. T. Miller Maguire, there are nearly a dozen good books in French. As a supplement to these facts is the spectacle of the officers of the Guards telegraphing to Sir Thomas Lipton on the occasion of the defeat of his Shamrock II., "Hard luck. Be of good cheer. Brigade of Guards wish you every success." This is not the foolish enthusiasm of one or two subalterns, it is collective. They followed that yacht race with emotion! as a really important thing to them. No doubt the whole mess was in a state of extreme excitement. How can capable and active men be expected to live and work between this upper and that nether millstone? The British army not only does not attract ambitious, energetic men, it repels them. I must confess that I see no hope either in the rulers, the traditions, or the manhood of the British regular army, to forecast its escape from the bog of ignorance and negligence in which it wallows. Far better than any of projected reforms would it be to let the existing army severely alone, to cease to recruit for it, to retain (at the expense of its officers, assisted perhaps by subscriptions from ascendant people like Sir Thomas Lipton) its messes, its uniforms, its games, bands, entertainments, and splendid memories as an appendage of the Court, and to create, in absolute independence of it, battalions and batteries of efficient professional soldiers, without social prestige or social distinctions, without bands, dress uniforms, colours, chaplains or honorary colonels, and to embody these as a real marching army perpetually *en route* throughout the empire–a reading, thinking, experimenting army under an absolutely distinct war office, with its own colleges, *dépôts,* and training-camps perpetually ready for war. I cannot help but think that, if a hint were taken from the *Turbinia* syndicate, a few enterprising persons of means and intelligence might do much by private experiment to supplement and replace the existing state of affairs.

new social order of efficients, only very painfully and slowly, amidst these growing and yet disintegrating masses. And on the other hand we have the warlike drift of such a social body, the inevitable intensification of international animosities in such a body, the absolute determination evident in the scheme of things to smash such a body, to smash it just as far as it is such a body, under the hammer of war, that must finally bring about rapidly and under pressure the same result as that to which the peaceful evolution slowly tends. While we are as yet only thinking of a physiological struggle, of complex reactions and slow absorptions, comes War with the surgeon's knife. War comes to simplify the issue and line out the thing with knifelike cuts.

The law that dominates the future is glaringly plain. A people must develop and consolidate its educated efficient classes or be beaten in war and give way upon all points where its interests conflict with the interests of more capable people. It must foster and accelerate that natural segregation, which has been discussed in the third and fourth chapters of these "Anticipations," or perish. The war of the coming time will really be won in schools and colleges and universities, wherever men write and read and talk together. The nation that produces in the near future the largest proportional development of educated and intelligent engineers and agriculturists, of doctors, schoolmasters, professional soldiers, and intellectually active people of all sorts; the nation that most resolutely picks over, educates, sterilises, exports, or poisons its People of the Abyss; the nation that succeeds most subtly in checking gambling and the moral decay of women and homes that gambling inevitably entails; the nation that by wise interventions, death-duties and the like, contrives to expropriate and extinguish incompetent rich families while leaving individual ambitions free; the nation, in a word, that turns the greatest proportion of its irresponsible adiposity into social muscle, will certainly be the nation that will be the most powerful in warfare as in peace, will certainly be the ascendant or dominant nation before the year 2000. In the long run no heroism and no accidents can alter that. No flag-waving, no patriotic leagues, no visiting of essentially petty imperial personages hither and thither, no smashing of the windows of outspoken people nor seizures of papers and books, will arrest the march of national defeat. And this issue is already so plain and simple, the alternatives are becoming so pitilessly clear, that even in the stupidest court and the stupidest constituencies, it must presently begin in some dim way to be felt. A time will come when so many people will see this issue clearly that it will gravely affect political and social life. The patriotic party–the particular gang, that is, of lawyers, brewers, landlords, and railway directors that wishes to be dominant–will be forced to become an efficient party in profession at least, will be forced to stimulate and organise that educational and social development that may at last even bring patriotism under control. The rulers of the grey, the democratic politician and the democratic monarch, will be obliged year by year by the very nature of things to promote the segregation of colours within the grey, to foster the power that will finally supersede democracy and monarchy altogether, the power of the scientifically educated, disciplined specialist, and that finally is the power of sanity, the power of the thing that is provably right. It may be delayed, but it cannot be defeated; in the end it must arrive–if not to-day and among our people, then to-morrow and among another people, who will triumph in our overthrow. This is the lesson that must be learnt, that some tongue and kindred of the coming time must inevitably learn. But what tongue it will be, and what kindred that will first attain this new development, opens far more complex and far less certain issues than any we have hitherto considered.

VII. THE CONFLICT OF LANGUAGES

WE have brought together thus far in these Anticipations the material for the picture of a human community somewhere towards the year 2000. We have imagined its roads, the type and appearance of its homes, its social developments, its internal struggle for organisation; we have speculated upon its moral and æsthetic condition, read its newspaper, made an advanced criticism upon the lack of universality in its literature, and attempted to imagine it at war. We have decided in particular that unlike the civilised community of the immediate past which lived either in sharply-defined towns or agriculturally over a wide country, this population will be distributed in a quite different way, a little more thickly over vast urban regions and a little less thickly over less attractive or less convenient or less industrial parts of the world. And implicit in all that has been written there has appeared an unavoidable assumption that the coming community will be vast, something geographically more extensive than most, and geographically different from almost all existing communities, that the outline its creative forces will draw not only does not coincide with existing political centres and boundaries, but will be more often than not in direct conflict with them, uniting areas that are separated and separating areas that are united, grouping here half a dozen tongues and peoples together and there tearing apart homogeneous bodies and distributing the fragments among separate groups. And it will now be well to inquire a little into the general causes of these existing divisions, the political boundaries of to-day, and the still older contours of language and race.

It is first to be remarked that each of these sets of boundaries is superposed, as it were, on the older sets. The race areas, for example, which are now not traceable in Europe at all must have represented old regions of separation; the language areas, which have little or no essential relation to racial distribution, have also given way long since to the newer forces that have united and consolidated nations. And the still newer forces that have united and separated the nineteenth century states have been, and in many cases are still, in manifest conflict with "national" ideas.

Now, in the original separation of human races, in the subsequent differentiation and spread of languages, in the separation of men into nationalities, and in the union and splitting of states and empires, we have to deal essentially with the fluctuating manifestations of the same fundamental shaping factor which will determine the distribution of urban districts in the coming years. Every boundary of the ethnographical, linguistic, political, and commercial map–as a little consideration will show–has indeed been traced in the first place by the means of transit, under the compulsion of geographical contours.

There are evident in Europe four or five or more very distinct racial types, and since the methods and rewards of barbaric warfare and the nature of the chief chattels of barbaric trade have always been diametrically opposed to racial purity, their original separation could only have gone on through such an entire lack of communication as prevented either trade or warfare between the bulk of the differentiating bodies. These original racial types are now inextricably mingled. Unobservant, over-scholarly people talk or write in the profoundest manner about a Teutonic race and a Keltic race, and institute all sorts of curious contrasts between these phantoms, but these are not races at all, if physical characteristics have anything to do with race. The Dane, the Bavarian, the Prussian, the Frieslander, the Wessex peasant, the Kentish man, the Virginian, the man from New Jersey, the Norwegian, the Swede, and the Transvaal Boer, are generalised about, for example, as Teutonic, while the short, dark, cunning sort of Welshman, the tall and generous Highlander, the miscellaneous Irish, the square-headed Breton, and any sort of Cornwall peasant are Kelts within the meaning of this oil-lamp anthropology.*

*Under the intoxication of the Keltic Renascence the most diverse sorts of human beings have foregathered and met face to face, and been photographed Pan-Keltically, and have no doubt gloated over these collective photographs, without any of them realising, it seems, what a miscellaneous thing the Keltic

People who believe in this sort of thing are not the sort of people that one attempts to convert by a set argument. One need only say the thing is not so; there is no Teutonic race, and there never has been; there is no Keltic race, and there never has been. No one has ever proved or attempted to prove the existence of such races, the thing has always been assumed; they are dogmas with nothing but questionable authority behind them, and the onus of proof rests on the believer. This nonsense about Keltic and Teutonic is no more science than Lombroso's extraordinary assertions about criminals, or palmistry, or the development of religion from a solar myth. Indisputably there are several races intermingled in the European populations–I am inclined to suspect the primitive European races may be found to be so distinct as to resist confusion and pamnyxia through hybridisation–but there is no inkling of a satisfactory analysis yet that will discriminate what these races were and define them in terms of physical and moral character. The fact remains there is no such thing as a racially pure and homogenous community in Europe distinct from other communities. Even among the Jews, according to Erckert and Chantre and J. Jacobs, there are markedly divergent types, there may have been two original elements and there have been extensive local intermixtures.

Long before the beginnings of history, while even language was in its first beginnings–indeed as another aspect of the same process as the beginning of language–the first complete isolations that established race were breaking down again, the little pools of race were running together into less homogeneous lagoons and marshes of humanity, the first paths were being worn–war paths for the most part. Still differentiation would be largely at work. Without frequent intercourse, frequent interchange of women as the great factor in that intercourse, the tribes and bands of mankind would still go on separating, would develop dialectic and customary, if not physical and moral differences. It was no longer a case of pools perhaps, but they were still in lakes. There were as yet no open seas of mankind. With advancing civilisation, with iron weapons and war discipline, with established paths and a social rule and presently with the coming of the horse, what one might call the areas of assimilation would increase in size. A stage would be reached when the only checks to transit of a sufficiently convenient sort to keep language uniform would be

race must be. There is nothing that may or may not be a Kelt, and I know, for example, professional Kelts who are, so far as face, manners, accents, morals, and ideals go, indistinguishable from other people who are, I am told, indisputably Assyroid Jews.

the sea or mountains or a broad river or–pure distance. And presently the rules of the game, so to speak, would be further altered and the unifications and isolations that were establishing themselves upset altogether and brought into novel conflict by the beginnings of navigation, whereby an impassable barrier became a highway.

The commencement of actual European history coincides with the closing phases of what was probably a very long period of a foot and (occasionally) horseback state of communications; the adjustments so arrived at being already in an early state of rearrangement through the advent of the ship. The communities of Europe were still for the larger part small isolated tribes and kingdoms, such kingdoms as a mainly pedestrian militia, or at anyrate a militia without transport, and drawn from (and soon drawn home again by) agricultural work, might hold together. The increase of transit facilities between such communities, by the development of shipping and the invention of the wheel and the made road, spelt increased trade perhaps for a time, but very speedily a more extensive form of war, and in the end either the wearing away of differences and union, or conquest. Man is the creature of a struggle for existence, incurably egoistic and aggressive. Convince him of the gospel of self-abnegation even, and he instantly becomes its zealous missionary, taking great credit that his expedients to ram it into the minds of his fellow-creatures do not include physical force–and if that is not self-abnegation, he asks, what is? So he has been, and so he is likely to remain. Not to be so, is to die of abnegation and extinguish the type. Improvement in transit between communities formerly for all practical purposes isolated, means, therefore, and always has meant, and I imagine, always will mean, that now they can get at one another. And they do. They inter-breed and fight, physically, mentally, and spiritually. Unless Providence is belied in His works that is what they are meant to do.

A third invention which, though not a means of transit like the wheeled vehicle and the ship, was yet a means of communication, rendered still larger political reactions possible, and that was the development of systems of writing. The first empires and some sort of written speech arose together. Just as a kingdom, as distinguished from a mere tribal group of villages, is almost impossible without horses, so is an empire without writing and post-roads. The history of the whole world for three thousand years is the history of a unity larger than the small kingdom of the Heptarchy type, endeavouring to establish itself under the stress of these discoveries of horse-traffic and shipping and the written word, the history, that is, of the consequences of the partial shattering of the barriers that had been effectual enough to prevent the fusion of more than tribal communities through all the long ages before the dawn of history.

East of the Gobi Pamir barrier there has slowly grown up under these new conditions the Chinese system. West and north of the Sahara Gobi barrier of deserts and mountains, the extraordinarily strong and spacious conceptions of the Romans succeeded in dominating the world, and do, indeed, in a sort of mutilated way, by the powers of great words and wide ideas, in Cæsarism and Imperialism, in the titles of Czar, Kaiser, and Imperator, in Papal pretension and countless political devices, dominate it to this hour. For awhile these conceptions sustained a united and to a large extent organised empire over very much of this space. But at its stablest time, this union was no more than a political union, the spreading of a thin layer of Latin-speaking officials, of a thin network of roads and a very thin veneer indeed of customs and refinements, over the scarcely touched national masses. It checked, perhaps, but it nowhere succeeded in stopping the slow but inevitable differentiation of province from province and nation from nation. The forces of transit that permitted the Roman imperialism and its partial successors to establish wide ascendancies, were not sufficient to carry the resultant unity beyond the political stage. There was unity, but not unification. Tongues and writing ceased to be pure without ceasing to be distinct. Sympathies, religious and social practices, ran apart and rounded themselves off like drops of oil on water. Travel was restricted to the rulers and the troops and to a wealthy leisure class; commerce was for most of the constituent provinces of the empire a commerce in superficialities, and each province–except for Italy, which latterly became dependent on an over-seas food supply–was in all essential things autonomous, could have continued in existence, rulers and ruled, arts, luxuries, and refinements just as they stood, if all other lands and customs had been swept out of being. Local convulsions and revolutions, conquests and developments, occurred indeed, but though the stones were altered the mosaic remained, and the general size and character of its constituent pieces remained. So it was under the Romans, so it was in the eighteenth century, and so it would probably have remained as long as the post-road and the sailing-ship were the most rapid forms of transit within the reach of man. Wars and powers and princes came and went, that was all. Nothing was changed, there was only one state the more or less.

Even in the eighteenth century the process of real unification had effected so little, that not one of the larger kingdoms of Europe escaped a civil war–not a class war, but a really *internal* war–between one part of itself and another, in that hundred years. In spite of Rome's few centuries of unstable empire, internal wars, a perpetual struggle against finally triumphant disruption seemed to be the unavoidable destiny of every power that attempted to rule over a larger radius than at most a hundred miles.

So evident was this that many educated English persons thought then, and many who are not in the habit of analysing operating causes, still think to-day, that the wide diffusion of the English-speaking people is a mere preliminary to their political, social, and linguistic disruption–the eighteenth-century breach with the United States is made a precedent of, and the unification that followed the war of Union and the growing unification of Canada is overlooked–that linguistic differences, differences of custom, costume, prejudice, and the like, will finally make the Australian, the Canadian of English blood, the Virginian, and the English Africander, as incomprehensible and unsympathetic one to another as Spaniard and Englishman or Frenchman and German are now. On such a supposition all our current Imperialism is the most foolish defiance of the inevitable, the maddest waste of blood, treasure, and emotion that man ever made. So, indeed, it might be–so, indeed, I certainly think it would be–if it were not that the epoch of post-road and sailing-ship is at an end. We are in the beginning of a new time, with such forces of organisation and unification at work in mechanical traction, in the telephone and telegraph, in a whole wonderland of novel, space-destroying appliances, and in the correlated inevitable advance in practical education, as the world has never felt before.

The operation of these unifying forces is already to be very distinctly traced in the check, the arrest indeed, of any further differentiation in existing tongues, even in the most widely spread. In fact, it is more than an arrest even, the forces of differentiation have been driven back and an actual process of assimilation has set in. In England at the commencement of the nineteenth century the common man of Somerset and the common man of Yorkshire, the Sussex peasant, the Caithness cotter and the common Ulsterman, would have been almost incomprehensible to one another. They differed in accent, in idiom, and in their very names for things. They differed in their ideas about things. They were, in plain English, foreigners one to another. Now they differ only in accent, and even that is a dwindling difference. Their language has become ampler because now they read. They read books–or, at anyrate, they learn to read out of books–and certainly they read newspapers and those scrappy periodicals that people like bishops pretend to think so detrimental to the human mind, periodicals that it is cheaper to make at centres and uniformly, than locally in accordance with local needs. Since the newspaper cannot fit the locality, the locality has to broaden its mind to the newspaper, and to ideas acceptable in other localities. The word and the idiom of the literary language and the pronunciation suggested by its spelling tends to prevail over the local usage. And moreover there is a persistent mixing of peoples going on, migration in search of employment and so on, quite unprecedented before the railways came. Few people are content to remain in that locality and state of life "into which it has pleased God to call them." As a result, dialectic purity has vanished, dialects are rapidly vanishing, and novel differentiations are retarded or arrested altogether. Such novelties as do establish themselves in a locality are widely disseminated almost at once in books and periodicals.

A parallel arrest of dialectic separation has happened in France, in Italy, in Germany, and in the States. It is not a process peculiar to any one nation. It is simply an aspect of the general process that has arisen out of mechanical locomotion. The organisation of elementary education has no doubt been an important factor, but the essential influence working through this circumstance is the fact that paper is relatively cheap to type-setting, and both cheap to authorship–even the commonest sorts of authorship–and the wider the area a periodical or book serves the bigger, more attractive, and better it can be made for the same money. And clearly this process of assimilation will continue. Even local differences of accent seem likely to follow. The itinerant dramatic company, the itinerant preacher, the coming extension of telephones and the phonograph, which at any time in some application to correspondence or instruction may cease to be a toy, all these things attack, or threaten to attack, the weeds of differentiation before they can take root. . . .

And this process is not restricted to dialects merely. The native of a small country who knows no other language than the tongue of his country becomes increasingly at a disadvantage in comparison with the user of any of the three great languages of the Europeanised world. For his literature he depends on the scanty writers who are in his own case and write, or have written, in his own tongue. Necessarily they are few, because necessarily with a small public there can be only subsistence for a few. For his science he is in a worse case. His country can produce neither teachers nor discoverers to compare with the numbers of such workers in the larger areas, and it will neither pay them to write original matter for his instruction nor to translate what has been written in other tongues. The larger the number of people reading a tongue, the larger–other things being equal–will be not only the output of more or less original literature in that tongue, but also the more profitable and numerous will be translations of whatever has value in other tongues. Moreover, the larger the reading public in any language the cheaper will it be to supply copies of the desired work. In the matter of current intelligence the case of the speaker of the small language is still worse. His newspaper will need to be cheaply served, his home intelligence will be cut and restricted, his foreign news belated and second-hand. Moreover, to travel even a little distance or to conduct anything but the smallest business enterprise will be exceptionally inconvenient to him. The Englishman who knows no language but his own may travel well-nigh all over the world and everywhere meet someone who can speak his tongue. But what of the Welsh-speaking Welshman? What of the Basque and the Lithuanian who can speak only his mother tongue? Everywhere such a man is a foreigner and with all the foreigner's disadvantages. In most places he is for all practical purposes deaf and dumb.

The inducements to an Englishman, Frenchman, or German to become bi-lingual are great enough nowadays, but the inducements to a speaker of the smaller languages are rapidly approaching compulsion. He must do it in self-defence. To be an educated man in his own vernacular has become an impossibility, he must either become a mental subject of one of the greater languages or sink to the intellectual status of a peasant. But if our analysis of social development was correct the peasant of to-day will be represented to-morrow by the people of no account whatever, the classes of extinction, the People of the Abyss. If that analysis was correct, the essential nation will be all of educated men, that is to say, the essential nation will speak some dominant language or cease to exist, whatever its primordial tongue may have been. It will pass out of being and become a mere local area of the lower social stratum,–a Problem for the philanthropic amateur.

The action of the force of attraction of the great tongues is cumulative. It goes on, as bodies fall, with a steady acceleration. The more the great tongues prevail over the little languages the less will be the inducement to write and translate into these latter, the less the inducement to master them with any care or precision. And so this attack upon the smaller tongues, this gravitation of those who are born to speak them, towards the great languages, is not only to be seen going on in the case of such languages as Flemish, Welsh, or Basque, but even in the case of Norwegian and of such a great and noble tongue as the Italian, I am afraid that the trend of things makes for a similar suppression. All over Italy is the French newspaper and the French book, French wins its way more and more there, as English, I understand, is doing in Norway, and English and German in Holland. And in the coming years when the reading public will, in the case of the Western nations, be practically the whole functional population, when travel will be more extensive and abundant, and the interchange of printed matter still cheaper and swifter–and above all with the spread of the telephone–the process of subtle, bloodless, unpremeditated annexation will conceivably progress much more rapidly even than it does at present. The Twentieth Century will see the effectual crowding out of most of the weaker languages–if not a positive crowding out, yet at least (as in Flanders) a supplementing of them by the super-position of one or other of a limited number of world-languages over the area in which each is spoken. This will go on not only in Europe, but with varying rates of progress and local eddies and interruptions over the whole world. Except in the special case of China and Japan, where there may be a unique development, the peoples of the world will escape from the wreckage of their too small and swamped and foundering social systems, only up the ladders of what one may call the aggregating tongues.

What will these aggregating world-languages be? If one has regard only to its extension during the nineteenth century one may easily incline to overrate the probabilities of English becoming the chief of these. But a great part of the vast extension of English that has occurred has been due to the rapid reproduction of originally English-speaking peoples, the emigration of foreigners into English-speaking countries in quantities too small to resist the contagion about them, and the compulsion due to the political and commercial preponderance of a people too illiterate to readily master strange tongues. None of these causes have any essential permanence. When one comes to look more closely into the question one is surprised to discover how slow the extension of English has been in the face of apparently far less convenient tongues. English still fails to replace the French language in French Canada, and its ascendency is doubtful to-day in South Africa, after nearly a century of British dominion. It has none of the contagious quality of French, and the small class that monopolises the direction of British affairs, and probably will monopolise it yet for several decades, has never displayed any great zeal to propagate its use. Of the few ideas possessed by the British governing class, the destruction and discouragement of schools and colleges, is, unfortunately, one of the chief, and there is an absolute incapacity to understand the political significance of the language question. The Hindoo who is at pains to learn and use English encounters something uncommonly like hatred disguised in a facetious form. He will certainly read little about himself in English that is not grossly contemptuous, to reward him for his labour. The possibilities that have existed, and that do still in a dwindling degree exist, for resolute statesmen to make English the common language of communication for all Asia south and east of the Himalayas, will have to develop of their own force or dwindle and pass away. They may quite probably pass away. There is no sign that either the English or the Americans have a sufficient sense of the importance of linguistic predominance in the future of their race to interfere with natural processes in this matter for many years to come.

Among peoples not actually subject to British or American rule, and who are neither waiters nor commercial travellers, the inducements to learn English, rather than French or German, do not increase. If our initial assumptions are right, the decisive factor in this matter is the amount of science and thought the acquisition of a language will afford the man who learns it. It becomes, therefore, a fact of very great significance that the actual number of books published in English is less than that in French or German, and that the proportion of serious books is very greatly less. A large proportion of English books are novels adapted to the minds of women, or of boys and superannuated business men, stories designed rather to allay than stimulate thought—they are the only books, indeed, that are profitable to publisher and author alike. In this connection they do not count, however; no foreigner is likely to learn English for the pleasure of reading Miss Marie Corelli in the original, or of drinking untranslatable elements from *The Helmet of Navarre*. The present conditions of book production for the English reading public offer no hope of any immediate change in this respect. There is neither honour nor reward—there is not even food or shelter—for the American or Englishman who devotes a year or so of his life to the adequate treatment of any spacious question, and so small is the English reading pubic with any special interest in science, that a great number of important foreign scientific works are never translated into English at all. Such interesting compilations as Bloch's work on war, for example, must be read in French; in English only a brief summary of his results is to be obtained, under a sensational heading.* Schopenhauer again is only to be got quite stupidly Bowdlerised, explained, and "selected" in English. Many translations that are made into English are made only to sell, they are too often the work of sweated women and girls–very often quite without any special knowledge of the matter they translate–they are difficult to read and untrustworthy to quote. The production of books in English, except the author be a wealthy amateur, rests finally upon the publishers, and publishers to-day stand a little lower than ordinary tradesmen in not caring at all whether the goods they sell are good or bad. Unusual books, they allege–and all good books are unusual–are "difficult to handle," and the author must pay the fine–amounting, more often than not, to the greater portion of his interest in the book. There is no criticism to control the advertising enterprises of publishers and authors, and no sufficiently intelligent reading public has differentiated out of the confusion to encourage attempts at critical discrimination. The organs on the great professions and technical trades are as yet not alive to the part their readers must play in the public life of the future, and ignore all but strictly technical publications. A bastard criticism, written in many cases by publishers' employees, a criticism having a very direct relation to the advertisement-columns, distributes praise and blame in the periodic press. There is no body of great men either in England or America, no intelligence in the British Court, that might by any form of recognition compensate the philosophical or scientific writer for poverty and popular neglect. The more powerful a man's intelligence the more distinctly he must see that to devote himself to increase the scientific or philosophical wealth of the English tongue will be to sacrifice comfort, the respect of the bulk of his contemporaries, and all the most delightful things of life, for the barren reward of a not very certain righteous self-applause. By brewing and dealing in tied houses,** or by selling pork and tea, or by stock-jobbing

**Is War Now Impossible?* and see also footnote, p. 119.

**It is entirely for their wealth that brewers have been ennobled in England, never because of their services as captains of a great industry. Indeed, these services have been typically poor. While these men were earning their peerages by the sort of proceedings that do secure men peerages under the British Crown, the German brewers were developing the art and science of brewing and by pandering with the profits so obtained to the pleasures of the established great, a man of energy may hope to rise to a pitch of public honour and popularity immeasurably in excess of anything attainable through the most splendid intellectual performances. Heaven forbid I should overrate public honours and the company of princes! But it is not always delightful to be splashed by the wheels of cabs. Always before there has been at least a convention that the Court of this country, and its aristocracy, were radiant centres of moral and intellectual influence, that they did to some extent check and correct the judgments of the cab-rank and the beer-house. But the British Crown of to-day, so far as it exists for science and literature at all, exists mainly to repudiate the claims of intellectual performance to public respect.

These things, if they were merely the grievances of the study, might very well rest there. But they must be recognised here because the intellectual decline of the published literature of the English language–using the word to cover all sorts of books–involves finally the decline of the language and of all the spacious political possibilities that go with the wide extension of a language. Conceivably, if in the coming years a deliberate attempt were made to provide sound instruction in English to all who sought it, and to all within the control of English-speaking Governments, if honour and emolument were given to literary men instead of being left to them to most indelicately take, and if the present sordid trade of publishing were so lifted as to bring the whole literature, the whole science, and all the contemporary thought of the world–not some selection of the world's literature, not some obsolete Encyclopædia sold meanly and basely to choke hungry minds, but a real publication of all that has been and is being done–within the reach of each man's need and desire who had the franchise of the tongue, then by the year 2000 I would prophesy that the whole functional body of human society would read, and perhaps even write and speak, our language. And not only that, but it might be the prevalent and everyday language of Scandinavia and Denmark and Holland, of all Africa, all North America, of the Pacific coasts of Asia and of India, the universal international language, and in a fair way to be the universal language of mankind. But such an enterprise demands a resolve and intelligence beyond all the immediate signs of the times; it implies a veritable renascence of intellectual

with remarkable energy and success. The Germans and Bohemians can now make light beers that the English brewers cannot even imitate; they are exporting beer to England in steadily increasing volume.

life among the English-speaking peoples. The probabilities of such a renascence will be more conveniently discussed at a later stage, when we attempt to draw the broad outline of the struggle for world-wide ascendency that the coming years will see. But here it is clear that upon the probability of such a renascence depends the extension of the language, and not only that, but the preservation of that military and naval efficiency upon which, in this world of resolute aggression, the existence of the English-speaking communities finally depends.

French and German will certainly be aggregating languages during the great portion of the coming years. Of the two I am inclined to think French will spread further than German. There is a disposition in the world, which the French share, to grossly undervalue the prospects of all things French, derived, so far as I can gather, from the facts that the French were beaten by the Germans in 1870, and that they do not breed with the *abandon* of rabbits or negroes. These are considerations that affect the dissemination of French very little. The French reading public is something different and very much larger than the existing French political system. The number of books published in French is greater than that published in English; there is a critical reception for a work published in French that is one of the few things worth a writer's having, and the French translators are the most alert and efficient in the world. One has only to see a Parisian bookshop, and to recall an English one, to realise the as yet unattainable standing of French. The serried ranks of lemon-coloured volumes in the former have the whole range of human thought and interest; there are no taboos and no limits, you have everything up and down the scale, from frank indecency to stark wisdom. It is a shop for men. I remember my amazement to discover three copies of a translation of that most wonderful book, *The Principles of Psychology* of Professor William James, in a shop in L'Avenue de l'Opera–three copies of a book that I have never seen anywhere in England outside my own house,–and I am an attentive student of bookshop windows! And the French books are all so pleasant in the page, and so cheap–they are for a people that buys to read. One thinks of the English bookshop, with its gaudy reach-me-downs of gilded and embossed cover, its horribly printed novels still more horribly "illustrated," the exasperating pointless variety in the size and thickness of its books. The general effect of the English book is that it is something sold by a dealer in *bric-à-brac,* honestly sorry the thing is a book, but who has done *his* best to remedy it, anyhow! And all the English shopful is either brand new fiction or illustrated travel (of *'Buns with the Grand Lama'* type), or gilded versions of the classics of past times done up to give away. While the French bookshop reeks of contemporary intellectual life!

These things count for French as against English now, and they will count for infinitely more in the coming years. And over German also French has many advantages. In spite of the numerical preponderance of books published in Germany, it is doubtful if the German reader has quite such a catholic feast before him as the reader of French. There is a mass of German fiction probably as uninteresting to a foreigner as popular English and American romance. And German compared with French is an unattractive language; unmelodious, unwieldy, and cursed with a hideous and blinding lettering that the German is too patriotic to sacrifice. There has been in Germany a more powerful parallel to what one may call the "honest Saxon" movement among the English, that queer mental twist that moves men to call an otherwise undistinguished preface a "Foreword," and find a pleasurable advantage over their fellow-creatures in a familiarity with "eftsoons." This tendency in German has done much to arrest the simplification of idiom, and checked the development of new words of classical origin. In particular it has stood in the way of the international use of scientific terms. The Englishman, the Frenchman, and the Italian have a certain community of technical, scientific, and philosophical phraseology, and it is frequently easier for an Englishman with some special knowledge of his subject to read and appreciate a subtle and technical work in French, than it is for him to fully enter into the popular matter of the same tongue. Moreover, the technicalities of these peoples, being not so immediately and constantly brought into contrast and contact with their Latin or Greek roots as they would be if they were derived (as are so many "patriotic" German technicalities) from native roots, are free to qualify and develop a final meaning distinct from their original intention. In the growing and changing body of science this counts for much. The indigenous German technicality remains clumsy and compromised by its everyday relations, to the end of time it drags a lengthening chain of unsuitable associations, And the shade of meaning, the limited qualification, that a Frenchman or Englishman can attain with a mere twist of the sentence, the German must either abandon or laboriously overstate with some colossal wormcast of parenthesis. . . . Moreover, against the German tongue there are hostile frontiers, there are hostile people who fear German preponderance, and who have set their hearts against its use. In Roumania, and among the Slav, Bohemian, and Hungarian peoples, French attacks German in the flank, and has as clear a prospect of predominance.

These two tongues must inevitably come into keen conflict; they will perhaps fight their battle for the linguistic conquest of Europe, and perhaps of the world, in a great urban region that will arise about the Rhine. Politically this region lies now in six independent States, but economically it must become one in the next fifty years. It will almost certainly be the greatest urban region in all the world except that which will arise in the eastern States of North America, and that which may arise somewhere about Hankow. It will stretch from Lille to Kiel, it will drive extensions along the Rhine valley into Switzerland, and fling an arm along the Moldau to Prague, it will be the industrial capital of the old world. Paris will be its West End, and it will stretch a spider's-web of railways and great roads of the new sort over the whole continent. Even when the coal-field industries of the plain give place to the industrial application of mountain-born electricity, this great city region will remain, I believe, in its present position at the seaport end of the great plain of the Old World. Considerations of transit will keep it where it has grown, and electricity will be brought to it in mighty cables from the torrents of the central European mountain mass. Its westward port may be Bordeaux or Milford Haven, or even some port in the south-west of Ireland–unless, which is very unlikely, the velocity of secure sea-travel can be increased beyond that of land locomotion. I do not see how this great region is to unify itself without some linguistic compromise–the Germanisation of the French-speaking peoples by force is too ridiculous a suggestion to entertain. Almost inevitably with travel, with transport communications, with every condition of human convenience insisting upon it, formally or informally a bi-lingual compromise will come into operation, and to my mind at least the chances seem even that French will emerge on the upper hand. Unless, indeed, that great renascence of the English-speaking peoples should, after all, so overwhelmingly occur as to force this European city to be tri-lingual, and prepare the way by which the whole world may at last speak together in one tongue.

These are the aggregating tongues. I do not think that any other tongues than these are quite likely to hold their own in the coming time. Italian may flourish in the city of the Po valley, but only with French beside it. Spanish and Russian are mighty languages, but without a reading public how can they prevail, and what prospect of a reading public has either? They are, I believe, already judged. By A.D. 2000 all these languages will be tending more and more to be the second tongues of bi-lingual communities, with French, or English, or less probably German winning the upper hand.

But when one turns to China there are the strangest possibilities. It is in Eastern Asia alone that there seems to be any possibility of a synthesis sufficiently great to maintain itself, arising outside of, and independently of, the interlocked system of mechanically sustained societies that is developing out of mediæval Christendom. Throughout Eastern Asia there is still, no doubt, a vast wilderness of languages, but over them all rides the Chinese writing. And very strong–strong enough to be very gravely considered–is the possibility of that writing taking up an orthodox association of sounds, and becoming a world speech. The Japanese written language, the language of Japanese literature, tends to assimilate itself to Chinese, and fresh Chinese words and expressions are continually taking root in Japan. The Japanese are a people quite abnormal and incalculable, with a touch of romance, a conception of honour, a quality of imagination, and a clearness of intelligence that renders possible for them things inconceivable of any other existing nation. I may be the slave of perspective effects, but when I turn my mind from the pettifogging muddle of the English House of Commons, for example, that magnified vestry that is so proud of itself as a club–when I turn from that to this race of brave and smiling people, abruptly destiny begins drawing with a bolder hand. Suppose the Japanese were to make up their minds to accelerate whatever process of synthesis were possible in China! Suppose, after all, I am not the victim of atmospheric refraction, and they are, indeed, as gallant and bold and intelligent as my baseless conception of them would have them be! They would almost certainly find co-operative elements among the educated Chinese. . . . But this is no doubt the lesser probability. In front and rear of China the English language stands. It has the start of all other languages–the mechanical advantage–the position. And if only we, who think and write and translate and print and put forth, could make it worth the world's having!

VIII. THE LARGER SYNTHESIS

WE have seen that the essential process arising out of the growth of science and mechanism, and more particularly out of the still developing new facilities of locomotion and communication science has afforded, is the deliquescence of the social organisations of the past, and the synthesis of ampler and still ampler and more complicated and still more complicated social unities. The suggestion is powerful, the conclusion is hard to resist, that, through whatever disorders of danger and conflict, whatever centuries of misunderstanding and bloodshed, men may still have to pass, this process nevertheless aims finally, and will attain to the establishment of one world-state at peace within itself. In the economic sense, indeed, a world-state is already established. Even to-day we do all buy and sell in the same markets–albeit the owners of certain ancient rights levy their tolls here and there–and the Hindoo starves, the Italian feels the pinch, before the Germans or the English go short of bread. There is no real autonomy any more in the world, no simple right to an absolute independence such as formerly the Swiss could claim. The nations and boundaries of to-day do no more than mark claims to exemptions, privileges, and corners in the market–claims valid enough to those whose minds and souls are turned towards the past, but absurdities to those who look to the future as the end and justification of our present stresses. The claim to political liberty amounts, as a rule, to no more than the claim of a man to live in a parish without observing sanitary precautions or paying rates because he had an excellent great-grandfather. Against all these old isolations, these obsolescent particularisms, the forces of mechanical and scientific development fight, and fight irresistibly; and upon the general recognition of this conflict, upon the intelligence and courage with which its inflexible conditions are negotiated, depends very largely the amount of bloodshed and avoidable misery the coming years will hold.

The final attainment of this great synthesis, like the social deliquescence and reconstruction dealt with in the earlier of these anticipations, has an air of being a process independent of any collective or conscious will in man, as being the expression of a greater Will; it is working now, and may work out to its end vastly, and yet at times almost imperceptibly, as some huge secular movement in Nature, the raising of a continent, the crumbling of a mountain-chain, goes on to its appointed culmination. Or one may compare the process to a net that has surrounded, and that is drawn continually closer and closer upon, a great and varied multitude of men. We may cherish animosities, we may declare imperishable distances, we may plot and counter-plot, make war and "fight to a finish"; the net tightens for all that.

Already the need of some synthesis at least ampler than existing national organisations is so apparent in the world, that at least five spacious movements of coalescence exist to-day; there is the movement called Anglo-Saxonism, the allied but finally very different movement of British Imperialism, the Pan-Germanic movement, Pan-Slavism, and the conception of a great union of the "Latin" peoples. Under the outrageous treatment of the white peoples an idea of unifying the "Yellow" peoples is pretty certain to become audibly and visibly operative before many years. These are all deliberate and justifiable suggestions, and they all aim to sacrifice minor differences in order to link like to like in greater matters, and so secure, if not physical predominance in the world, at least an effective defensive strength for their racial, moral, customary, or linguistic differences against the aggressions of other possible coalescences. But these syntheses or other similar synthetic conceptions, if they do not contrive to establish a rational social unity by sanely negotiated unions, will be forced to fight for physical predominance in the world. The whole trend of forces in the world is against the preservation of *local* social systems, however greatly and spaciously conceived. Yet it is quite possible that several or all of the cultures that will arise out of the development of these Pan-this-and-that movements may in many of their features survive, as the culture of the Jews has survived, political obliteration, and may disseminate themselves, as the Jewish system has disseminated itself, over the whole world-city. Unity by no means involves homogeneity. The greater the social organism the more complex and varied its parts, the more intricate and varied the interplay of culture and breed and character within it.

It is doubtful if either the Latin or the Pan-Slavic idea contains the promise of any great political unification. The elements of the Latin synthesis are dispersed in South and Central America and about the Mediterranean basin in a way that offers no prospect of an economic unity between them. The best elements of the French people lie in the western portion of what must become the greatest urban region of the Old World, the Rhine-Netherlandish region; the interests of North Italy draw that region away from the Italy of Rome and the South towards the Swiss and South Germany, and the Spanish and Portuguese speaking halfbreeds of South America have not only their own coalescences to arrange, but they lie already under the political tutelage of the United States. Nowhere except in France and North Italy is there any prospect of such an intellectual and educational evolution as is necessary before a great scheme of unification can begin to take effect. And the difficulties in the way of the pan-Slavic dream are far graver. Its realisation is enormously hampered by the division of its languages, and the fact that in the Bohemian language, in Polish and in Russian, there exist distinct literatures, almost equally splendid in achievement, but equally insufficient in quantity and range to establish a claim to replace all other Slavonic dialects. Russia, which should form the central mass of this synthesis, stagnates, relatively to the Western states, under the rule of reactionary intelligences; it does not develop, and does not seem likely to develop, the merest beginnings of that great educated middle class, with which the future so enormously rests. The Russia of to-day is indeed very little more than a vast breeding-ground for an illiterate peasantry, and the forecasts of its future greatness entirely ignore that dwindling significance of mere numbers in warfare which is the clear and necessary consequence of mechanical advance. To a large extent, I believe, the Western Slavs will follow the Prussians and Lithuanians, and be incorporated in the urbanisation of Western Europe, and the remoter portions of Russia seem destined to become–are indeed becoming–Abyss, a wretched and disorderly Abyss that will not even be formidable to the armed and disciplined peoples of the new civilisation, the last quarter of the earth, perhaps, where a barbaric or absentee nobility will shadow the squalid and unhappy destinies of a multitude of hopeless and unmeaning lives.

To a certain extent, Russia may play the part of a vaster Ireland, in her failure to keep pace with the educational and economic progress of nations which have come into economic unity with her. She will be an Ireland without emigration, a place for famines. And while Russia delays to develop anything but a fecund orthodoxy and this simple peasant life, the grooves and channels are growing ever deeper along which the currents of trade, of intellectual and moral stimulus, must presently flow towards the West. I see no region where anything like the comparatively dense urban regions that are likely to arise about the Rhineland and over the eastern states of America, for example, can develop in Russia. With railways planned boldly, it would have been possible, it might still be possible, to make about Odessa a parallel to Chicago, but the existing railways run about Odessa as though Asia were unknown; and when at last the commercial awakening of what is now the Turkish Empire comes, the railway-lines will probably run, not north or south, but from the urban region of the more scientific central Europeans down to Constantinople. The long-route land communications in the future will become continually more swift and efficient than Baltic navigation, and it is unlikely, therefore, that St. Petersburg has any great possibilities of growth. It was founded by a man whose idea of the course of trade and civilisation was the sea wholly and solely, and in the future the sea must necessarily become more and more a last resort. With its spacious prospects, its architectural magnificence, its political quality, its desertion by the new commerce, and its terrible peasant hinterland, it may come about that a striking analogy between St. Petersburg and Dublin will finally appear.

So much for the Pan-Slavic synthesis. It seems improbable that it can prevail against the forces that make for the linguistic and economic annexation of the greater part of European Russia and of the minor Slavonic masses, to the great Western European urban region.

The political centre of gravity of Russia, in its resistance to these economic movements, is palpably shifting eastward even to-day, but that carries it away from the Central European synthesis only towards the vastly more enormous attracting centre of China. Politically the Russian Government may come to dominate China in the coming decades, but the reality beneath any such formal predominance will be the absorption of Russia beyond the range of the European pull by the synthesis of Eastern Asia. Neither the Russian literature nor the Russian language and writing, nor the Russian civilisation as a whole have the qualities to make them irresistible to the energetic and intelligent millions of the far East. The chances seem altogether against the existence of a great Slavonic power in the world at the beginning of the twenty-first century. They seem, at the first glance, to lie just as heavily in favour of an aggressive Pan-Germanic power struggling towards a great commanding position athwart Central Europe and Western Asia, and turning itself at last upon the defeated Slavonic disorder. There can be no doubt that at present the Germans, with the doubtful exception of the United States, have the most efficient middle class in the world, their rapid economic progress is to a very large extent, indeed, a triumph of intelligence, and their political and probably their military and naval services are still conducted with a capacity and breadth of view that find no parallel in the world. But the very efficiency of the German as a German to-day, and the habits and traditions of victory he has accumulated for nearly forty years, may prove in the end a very doubtful blessing to Europe as a whole, or even to his own grandchildren. Geographical contours, economic forces, the trend of invention and social development, point to a unification of all Western Europe, but they certainly do not point to its Germanisation. I have already given reasons for anticipating that the French language may not only hold its own, but prevail against German in Western Europe. And there are certain other obstacles in the way even of the union of indisputable Germans. One element in Germany's present efficiency must become more and more of an encumbrance as the years pass. The Germanic idea is deeply interwoven with the traditional Empire and with the martinet methods of the Prussian monarchy. The intellectual development of the Germans is defined to a very large extent by a court-directed officialdom. In many things that court is still inspired by the noble traditions of education and discipline that come from the days of German adversity, and the predominance of the Imperial will does, no doubt, give a unity of purpose to German policy and action that adds greatly to its efficacy. But for a capable ruler, even more than for a radiantly stupid monarch, the price a nation must finally pay is heavy. Most energetic and capable people are a little intolerant of unsympathetic capacity, are apt on the under side of their egotism to be jealous, assertive, and aggressive. In the present Empire of Germany there are no other great figures to balance the Imperial personage, and I do not see how other great figures are likely to arise. A great number of fine and capable persons must be failing to develop, failing to tell, under the shadow of this monarchy. There are certain limiting restrictions imposed upon Germans through the Imperial activity, that must finally be bad for the intellectual atmosphere which is Germany's ultimate strength. Discipline and education have carried Germany far; they are essential things, but an equally essential need for the coming time is a free play for men of initiative and imagination. Is Germany to her utmost possibility making capable men? That, after all, is the vital question, and not whether her policy is wise or foolish, or her commercial development inflated or sound. Or is Germany doing no more than cash the promises of earlier days?

After all, I do not see that she is in a greatly stronger position than was France in the early sixties, and, indeed, in many respects her present predominance is curiously analogous to that of the French Empire in those years. Death at any time may end the career of the present ruler of Germany–there is no certain insurance of one single life. This withdrawal would leave Germany organised entirely with reference to a Court, and there is no trustworthy guarantee as to the nature and opinions of the succeeding Royal Personality. Much has been done in the past of Germany, the infinitely less exacting past, by means of the tutor, the Chamberlain, the Chancellor, the wide-seeing power beyond the throne, who very unselfishly intrigues his monarch in the way that he should go. But that sort of thing is remarkably like writing a letter by means of a pen held in lazy tongs instead of the hand. A very easily imagined series of accidents may place the destinies of Germany in such lazy tongs again. When that occasion comes, will the new class of capable men on which we have convinced ourselves in these anticipations the future depends–will it be ready for its enlarged responsibilities, or will the flower of its possible members be in prison for *lèse majesté,* or naturalised Englishmen or naturalised Americans or troublesome privates under officers of indisputably aristocratic birth, or well-broken labourers, won "back to the land," under the auspices of an Agrarian League?

In another way the intensely monarchical and aristocratic organisation of the German Empire will stand in the way of the political synthesis of greater Germany. Indispensable factors in that synthesis will be Holland and Switzerland–little, advantageously situated peoples, saturated with ideas of personal freedom. One can imagine a German Swiss, at anyrate, merging himself in a great Pan-Germanic republican state, but to bow the knee to the God of His Imperial Majesty's Fathers will be an altogether more difficult exploit for a self-respecting man. . . .

Moreover, before Germany can unify to the East she must fight the Russian, and to unify to the West she must fight the French and perhaps the English, and she may have to fight a combination of these powers. I think the military strength of France is enormously underrated. Upon this matter M. Bloch should be read. Indisputably the French were beaten in 1870, indisputably they have fallen behind in their long struggle to maintain themselves equal with the English on the sea, but neither of these things efface the future of the French. The disasters of 1870 were probably of the utmost benefit to the altogether too sanguine French imagination. They cleared the French mind of the delusion that personal Imperialism is the way to do the desirable thing, a delusion many Germans (and, it would seem, a few queer Englishmen and still queerer Americans) entertain. The French have done much to demonstrate the possibility of a stable military republic. They have disposed of crown and court, and held themselves in order for thirty good years; they have dissociated their national life from any form of religious profession; they have contrived a freedom of thought and writing that, in spite of much conceit to the contrary, is quite impossible among the English-speaking peoples. I find no reason to doubt the implication of M. Bloch that on land to-day the French are relatively far stronger than they were in 1870, that the evolution of military expedients has been all in favour of the French character and intelligence, and that even a single-handed war between France and Germany to-day might have a very different issue from that former struggle. In such a conflict it will be Germany, and not France, that will have pawned her strength to the English-speaking peoples on the high seas. And France will not fight alone. She will fight for Switzerland or Luxembourg, or the mouth of the Rhine. She will fight with the gravity of remembered humiliations, with the whole awakened Slav race at the back of her antagonist, and very probably with the support of the English-speaking peoples.

It must be pointed out how strong seems the tendency of the German Empire to repeat the history of Holland upon a larger scale. While the Dutch poured out all their strength upon the seas, in a conflict with the English that at the utmost could give them only trade, they let the possibilities of a great Low German synthesis pass utterly out of being. (In those days Low Germany stretched to Arras and Douay.) They positively dragged the English into the number of their enemies. And to-day the Germans invade the sea with a threat and intention that will certainly create a countervailing American navy, fundamentally modify the policy of Great Britain, such as it is, and very possibly go far to effect the synthesis of the English-speaking peoples.

So involved, I do not see that the existing Germanic synthesis is likely to prevail in the close economic unity, the urban region that will arise in Western Europe. I imagine that the German Empire–that is, the organised expression of German aggression to-day–will be either shattered or weakened to the pitch of great compromises by a series of wars by land and sea; it will be forced to develop the autonomy of its rational middle class in the struggles that will render these compromises possible, and it will be finally not Imperial German ideas, but central European ideas possibly more akin to Swiss conceptions, a civilised republicanism finding its clearest expression in the French language, that will be established upon a bilingual basis throughout Western Europe, and increasingly predominant over the whole European mainland and the Mediterranean basin, as the twentieth century closes. The splendid dream of a Federal Europe, which opened the nineteenth century for France, may perhaps, after all, come to something like realisation at the opening of the twenty-first. But just how long these things take, just how easily or violently they are brought about, depends, after all, entirely upon the rise in general intelligence in Europe. An ignorant, a merely trained or a merely cultured people, will not understand these coalescences, will fondle old animosities and stage hatreds, and for such a people there must needs be disaster, forcible conformities and war. Europe will have her Irelands as well as her Scotlands, her Irelands of unforgettable wrongs, kicking, squalling, bawling most desolatingly, for nothing that anyone can understand. There will be great scope for the shareholding dilettanti, great opportunities for literary quacks, in "national" movements, language leagues, picturesque plotting, and the invention of such "national" costumes as the world has never seen. The cry of the little nations will go up to heaven, asserting the inalienable right of all little nations to sit down firmly in the middle of the high-road, in the midst of the thickening traffic, and with all their dear little toys about them, play and play–just as they used to play before the road had come. . . .

And while the great states of the continent of Europe are hammering down their obstructions of language and national tradition or raising the educational level above them until a working unity is possible, and while the reconstruction of Eastern Asia–whether that be under Russian, Japanese, English, or native China direction–struggles towards attainment, will there also be a great synthesis of the English-speaking peoples going on? I am inclined to believe that there will be such a synthesis, and that the head and centre of the new unity will be the great urban region that is developing between Chicago and the Atlantic, and which will lie mainly, but not entirely, south of the St. Lawrence. Inevitably, I think, that region must become the intellectual, political, and industrial centre of any permanent unification of the English-speaking states. There will, I believe, develop about that centre a great federation of white English-speaking peoples, a federation having America north of Mexico as its central mass (a federation that may conceivably include Scandinavia) and its federal government will sustain a common fleet, and protect or dominate or actually administer most or all of the non-white states of the present British Empire, and in addition much of the South and Middle Pacific, the East and West Indies, the rest of America, and the larger part of black Africa. Quite apart from the dominated races, such an English-speaking state should have by the century-end a practically homogeneous citizenship of at least a hundred million sound-bodied and educated and capable *men*. It should be the first of the three powers of the world, and it should face the organising syntheses of Europe and Eastern Asia with an intelligent sympathy. By the year 2000 all its common citizens should certainly be in touch with the thought of Continental Europe through the medium of French; its English language should be already rooting firmly through all the world beyond its confines, and its statesmanship should be preparing openly and surely, and discussing calmly with the public mind of the European, and probably of the Yellow state, the possible coalescences and conventions, the obliteration of custom-houses, the homologisation of laws and coinage and measures, and the mitigation of monopolies and special claims, by which the final peace of the world may be assured for ever. Such a synthesis, at anyrate, of the peoples now using the English tongue, I regard not only as a possible, but as a probable, thing. The positive obstacles to its achievement, great though they are, are yet trivial in comparison with the obstructions to that lesser European synthesis we have ventured to forecast. The greater obstacle is negative, it lies in the want of stimulus, in the lax prosperity of most of the constituent states of such a union. But such a stimulus, the renascence of Eastern Asia, or a great German fleet upon the ocean, may presently supply.

Now, all these three great coalescences, this shrivelling up and vanishing of boundary-lines, will be the outward and visible accompaniment of that inward and social reorganisation which it is the main object of these Anticipations to display. I have sought to show that in peace and war alike a process has been and is at work, a process with all the inevitableness and all the patience of a natural force, whereby the great swollen, shapeless, hypertrophied social mass of to-day must give birth at last to a naturally and informally organised, educated class, an unprecedented sort of people, a New Republic dominating the world. It will be none of our ostensible governments that will effect this great clearing up; it will be the mass of power and intelligence altogether outside the official state systems of to-day that will make this great clearance, a new social Hercules that will strangle the serpents of war and national animosity in his cradle.

Now, the more one descends from the open uplands of wide generalisation to the parallel jungle of particulars, the more dangerous does the road of prophesying become, yet nevertheless there may be some possibility of speculating how, in the case of the English-speaking synthesis at least, this effective New Republic may begin visibly to shape itself out and appear. It will appear first, I believe, as a conscious organisation of intelligent and quite possibly in some cases wealthy men, as a movement having distinct social and political aims, confessedly ignoring most of the existing apparatus of political control, or using it only as an incidental implement in the attainment of these aims. It will be very loosely organised in its earlier stages, a mere movement of a number of people in a certain direction, who will presently discover with a sort of surprise the common object towards which they are all moving.

Already there are some interesting aspects of public activity that, diverse though their aim may seem, do nevertheless serve to show the possible line of development of this New Republic in the coming time. For example, as a sort of preliminary sigh before the stirring of a larger movement, there are various Anglo-American movements and leagues to be noted. Associations for entertaining travelling samples of the American leisure class in guaranteed English country houses, for bringing them into momentary physical contact with real titled persons at lunches and dinners, and for having them collectively lectured by respectable English authors and divines, are no doubt trivial things enough;

but a snob sometimes shows how the wind blows better than a serious man. The Empire may catch the American as the soldier caught the Tartar. There is something very much more spacious than such things as this, latent in both the British and the American mind, and observable, for instance, in the altered tone of the Presses of both countries since the Venezuela Message and the Spanish-American War. Certain projects of a much ampler sort have already been put forward. An interesting proposal of an interchangeable citizenship, so that with a change of domicile an Englishman should have the chance of becoming a citizen of the United States, and an American a British citizen or a voter in an autonomous British colony, for example, has been made. Such schemes will, no doubt, become frequent, and will afford much scope for discussion in both countries during the next decade or so.* The American constitution and the British crown and constitution have to be modified or shelved at some stage in this synthesis, and for certain types of intelligence there could be no more attractive problem. Certain curious changes in the colonial point of view will occur as these discussions open out. The United States of America are rapidly taking, or having already taken, the ascendency in the iron and steel and electrical industries out of the hands of the British; they are developing a far ampler and more thorough system of higher scientific education than the British, and the spirit of efficiency percolating from their more efficient businesses is probably higher in their public services. These things render the transfer of the present mercantile and naval ascendency of Great Britain to the United States during the next two or three decades a very probable thing, and when this is accomplished the problem how far colonial loyalty is the fruit of Royal Visits and sporadic knighthoods, and how far it has relation to the existence of a predominant fleet, will be near its solution. An interesting point about such discussions as this, in which indeed in all probability the nascent consciousness of the New Republic will emerge, will be the solution this larger synthesis will offer to certain miserable difficulties of the present time. Government by the elect of the first families of Great Britain has in the last hundred years made Ireland and South Africa two open sores of irreconcilable wrong. These two English-speaking communities will never rest and never emerge from

*I foresee great scope for the ingenious persons who write so abundantly to the London evening papers upon etymological points, issues in heraldry, and the correct Union Jack, in the very pleasing topic of a possible Anglo-American flag (for use at first only on unofficial occasions).

wretchedness under the vacillating vote-catching incapacity of British Imperialism, and it is impossible that the British power, having embittered them, should ever dare to set them free. But within such an ampler synthesis as the New Republic will seek, these states could emerge to an equal fellowship that would take all the bitterness from their unforgettable past.

Another type of public activity which foreshadows an aspect under which the New Republic will emerge is to be found in the unofficial organisations that have come into existence in Great Britain to watch and criticise various public departments. There is, for example, the Navy League, a body of intelligent and active persons with a distinctly expert qualification which has intervened very effectively in naval control during the last few years. There is also at present a vast amount of disorganised but quite intelligent discontent with the tawdry futilities of army reform that occupy the War Office. It becomes apparent that there is no hope of a fully efficient and well-equipped official army under parliamentary government, and with that realisation there will naturally appear a disposition to seek some way to military efficiency, as far as is legally possible, outside War Office control. Already recruiting is falling off, it will probably fall off more and more as the patriotic emotions evoked by the Boer War fade away, and no trivial addition to pay or privilege will restore it. Elementary education has at last raised the intelligence of the British lower classes to a point when the prospect of fighting in distant lands under unsuitably educated British officers of means and gentility with a defective War Office equipment and inferior weapons has lost much of its romantic glamour. But an unofficial body that set itself to the establishment of a school of military science, to the sane organisation and criticism of military experiments in tactics and equipment, and to the raising for experimental purposes of volunteer companies and battalions, would find no lack of men. . . . What an unofficial syndicate of capable persons of the new sort may do in these matters has been shown in the case of the *Turbinia,* the germ of an absolute revolution in naval construction.

Such attempts at unofficial soldiering would be entirely in the spirit in which I believe the New Republic will emerge, but it is in another line of activity that the growing new consciousness will presently be much more distinctly apparent. It is increasingly evident that to organise and control public education is beyond the power of a democratic government. The meanly equipped and pretentiously conducted private schools of Great

Britain, staffed with ignorant and incapable young men, exist, on the other hand, to witness that public education is no matter to be left to merely commercial enterprise working upon parental ignorance and social prejudice. The necessary condition to the effective development of the New Republic is a universally accessible, spacious, and varied educational system working in an atmosphere of efficient criticism and general intellectual activity. Schools alone are of no avail, universities are merely dens of the higher cramming, unless the schoolmasters and schoolmistresses and lecturers are in touch with and under the light of an abundant, contemporary, and fully adult intellectuality. At present, in Great Britain at least, the head-masters entrusted with the education of the bulk of the influential men of the next decades are conspicuously second-rate men, forced and etiolated creatures, scholarship boys manured with annotated editions, and brought up under and protected from all current illumination by the kalepot of the Thirty-nine Articles. Many of them are less capable teachers and even less intelligent men than many Board School teachers. There is, however, urgent need of an absolutely new type of school–a school that shall be, at least, so skilfully conducted as to supply the necessary training in mathematics, dialectics, languages, and drawing, and the necessary knowledge of science, without either consuming all the leisure of the boy or destroying his individuality, as it is destroyed by the ignorant and pretentious blunderers of to-day; and there is an equally manifest need of a new type of University, something other than a happy fastness for those precociously brilliant creatures–creatures whose brilliance is too often the hectic indication of a constitutional unsoundness of mind–who can "get in" before the portcullis of the nineteenth birthday falls. These new educational elements may either grow slowly through the steady and painful pressure of remorseless facts, or, as the effort to evoke the New Republic becomes more conscious and deliberate, they may be rapidly brought into being by the conscious endeavours of capable men. Assuredly they will never be developed by the wisdom of the governments of the grey. It may be pointed out that in an individual and disorganised way a growing sense of such needs is already displayed. Such great business managers as Mr. Andrew Carnegie, for example, and many other of the wealthier efficients of the United States of America, are displaying a strong disinclination to found families of functionless shareholders, and a strong disposition to contribute, by means of colleges, libraries, and splendid foundations, to the future of the whole English-speaking world. Of course, Mr. Carnegie is not an educational specialist, and his good intentions will be largely exploited by the energetic mediocrities who control our educational affairs. But it is the intention that concerns us now, and not the precise method or effect. Indisputably these rich Americans are at a fundamentally important work in these endowments, and as indisputably many of their successors–I do not mean the heirs to their private wealth, but the men of the same type who will play their *rôle* in the coming years–will carry on this spacious work with a wider prospect and a clearer common understanding.

The establishment of modern and efficient schools is alone not sufficient for the intellectual needs of the coming time. The school and university are merely the preparation for the life of mental activity in which the citizen of the coming state will live. The three years of university and a lifetime of garrulous stagnation which constitutes the mind's history of many a public school-master, for example, and most of the clergy to-day, will be impossible under the new needs. The old-fashioned university, secure in its omniscience, merely taught; the university of the coming time will, as its larger function, criticise and learn. It will be organised for research–for the criticism, that is, of thought and nature. And a subtler and a greater task before those who will presently swear allegiance to the New Republic is to aid and stimulate that process of sound adult mental activity which is the cardinal element in human life. After all, in spite of the pretentious impostors who trade upon the claim, literature, contemporary literature, is the breath of civilised life, and those who sincerely think and write the salt of the social body. To mumble over the past, to live on the classics, however splendid, is senility. The New Republic, therefore, will sustain its authors. In the past the author lived within the limits of his patron's susceptibility, and led the world, so far as he did lead it, from that cage. In the present he lives within the limits of a particularly distressful and ill-managed market. He must please and interest the public before he may reason with it, and even to reach the public ear involves other assiduities than writing. To write one's best is surely sufficient work for a man, but unless the author is prepared to add to his literary toil the correspondence and alert activity of a business man, he may find that no measure of acceptance will save him from a mysterious poverty. Publishing has become a trade, differing only from the trade in pork or butter in the tradesman's careless bookkeeping and his professed indifference to the quality of his goods. But unless the whole mass of argument in these Anticipations is false, publishing is as much, or even more, of a public concern than education, and as little to be properly discharged by private men working for profit. On the other hand, it is not to be undertaken by a government of the grey, for a confusion cannot undertake to clarify itself: it is an activity in which the New Republic will necessarily engage.

The men of the New Republic will be intelligently critical men, and they will have the courage of their critical conclusions. For the sake of the English tongue, for the sake of the English peoples, they will set themselves to put temptingly within the reach of all readers of the tongue, and all possible readers of the tongue, an abundance of living literature. They will endeavour to shape great publishing trusts and associations that will have the same relation to the publishing office of to-day that a medical association has to a patent-medicine dealer. They will not only publish, but sell; their efficient book-shops, their efficient system of book-distribution will replace the present haphazard dealings of quite illiterate persons under whose shadows people in the provinces live.* If one of these publishing groups decides that a book, new or old, is of value to the public mind, I conceive the copyright will be secured and the book produced all over the world in every variety of form and price that seems necessary to its exhaustive sale. Moreover, these publishing associations will sustain spaciously conceived organs of opinion and criticism, which will begin by being patiently and persistently good, and so develop into power. And the more distinctly the New Republic emerges, the less danger there will be of these associations being allowed to outlive their service in a state of ossified authority. New groups of men and new phases of thought will organise their publishing associations as children learn to talk.**

*In a large town like Folkestone, for example, it is practically impossible to buy any book but a "boomed" novel unless one has ascertained the names of the author, the book, the edition, and the publisher. There is no index in existence kept up to date that supplies these particulars. If, for example, one wants–as I want (1) to read all that I have not read of the work of a Mr. Frank Stockton, (2) to read a book of essays by Professor Ray Lankaster the title of which I have forgotten, and (3) to buy the most convenient edition of the works of Swift, one has to continue wanting until the British Museum Library chances to get in one's way. The book-selling trade supplies no information at all on these points.

**One of the least satisfactory features of the intellectual atmosphere of the present time is the absence of good controversy. To follow closely an honest and subtle controversy, and to have arrived at a definite opinion upon some general question of real and practical interest and complicated reference, is assuredly the most educational exercise in the world–I would go so far as to say that no person is completely educated who has not done as much. The

And while the New Republic is thus developing its idea of itself and organising its mind, it will also be growing out of the confused and intricate businesses and undertakings and public services of the present time, into a recognisable material body. The synthetic process that is going on in the case of many of the larger of the businesses of the world, that formation of Trusts that

memorable discussions in which Huxley figured, for example, were extraordinarily stimulating. We lack that sort of thing now. A great number of people are expressing conflicting opinions upon all sorts of things, but there is a quite remarkable shirking of plain issues of debate. There is no answering back. There is much indirect answering, depreciation of the adversary, attempts to limit his publicity, restatements of the opposing opinion in a new way, but no conflict in the lists. We no longer fight obnoxious views, but assassinate them. From first to last, for example, there has been no honest discussion of the fundamental issues in the Boer War. Something may be due to the multiplication of magazine and newspapers, and the confusion of opinions that has scattered the controversy-following public. It is much to be regretted that the laws of copyright and the methods of publication stand in the way of annotated editions of works of current controversial value. For example, Mr. Andrew Lang has assailed the new edition of the "Golden Bough." His criticisms, which are, no doubt, very shrewd and penetrating, ought to be accessible with the text he criticises. Yet numerous people will read his comments who will never read the "Golden Bough"; they will accept his dinted sword as proof of the slaughter of Mr. Fraser, and many will read the "Golden Bough" and never hear of Mr. Lang's comments. Why should it be so hopeless to suggest an edition of the "Golden Bough" with footnotes by Mr. Lang and Mr. Fraser's replies? There are all sorts of books to which Mr. Lang might add footnotes with infinite benefit to everyone. Mr. Mallock, again, is going to explain how Science and Religion stand at the present time. If only someone would explain in the margin how Mr. Mallock stands, the thing would be complete. Such a book, again, as these "Anticipations" would stand a vast amount of controversial footnoting. It bristles with pegs for discussion–vacant pegs; it is written to provoke. I hope that some publisher, sooner or later, will do something of this kind, and will give us not only the text of an author's work, but a series of footnotes and appendices by reputable antagonists. The experiment, well handled, might prove successful enough to start a fashion–a very beneficial fashion for authors and readers alike. People would write twice as carefully and twice as clearly with that possible second edition (with footnotes by X and Y) in view. Imagine "The Impregnable Rock of Holy Scripture" as it might have been edited by the late Professor Huxley; Froude's edition of the "Grammar of Assent"; Mr. G. B. Shaw's edition of the works of Mr. Lecky; or the criticism of art and life of Ruskin,–the "Beauties of Ruskin" annotated by Mr. Whistler and carefully prepared for the press by Professor William James. Like the tomato and the cucumber, every book would carry its antidote wrapped about it. Impossible, you say. But is it? Or is it only unprecedented? If novelists will consent to the illustration of their stories by artists whose chief aim appears to be to contradict their statements, I do not see why controversial writers who believe their opinions are correct should object to the checking of their facts and logic by persons with a different way of thinking. Why should not men of opposite opinions collaborate in their discussion?

bulks so large in American discussion, is of the utmost significance in this connection. Conceivably the first impulse to form Trusts came from a mere desire to control competition and economise working-expenses, but even in its very first stages this process of coalescence has passed out of the region of commercial operations into that of public affairs. The Trust develops into the organisation under men far more capable than any sort of public officials, of entire industries, of entire departments of public life, quite outside the ostensible democratic government system altogether. The whole apparatus of communications, which we have seen to be of such primary importance in the making of the future, promises to pass, in the case of the United States at least, out of the region of scramble into the domain of deliberate control. Even to-day the Trusts are taking over quite consciously the most vital national matters. The American iron and steel industries have been drawn together and developed in a manner that is a necessary preliminary to the capture of the empire of the seas. That end is declaredly within the vista of these operations, within their initial design. These things are not the work of dividend-hunting imbeciles, but of men who regard wealth as a convention, as a means to spacious material ends. There is an animated little paper published in Los Angeles in the interests of Mr. Wilshire, which bears upon its forefront the maxim, "Let the Nation own the Trusts." Well, under their mantle of property, the Trusts grow continually more elaborate and efficient machines of production and public service, while the formal nation chooses its bosses and buttons and reads its illustrated press. I must confess I do not see the negro and the poor Irishman and all the emigrant sweepings of Europe, which constitute the bulk of the American Abyss, uniting to form that great Socialist party of which Mr. Wilshire dreams, and with a little demonstrating and balloting taking over the foundry and the electrical works, the engine-shed and the signal-box, from the capable men in charge. But that a confluent system of Trust-owned business organisms, and of Universities and re-organised military and naval services may presently discover an essential unity of purpose, presently begin thinking a literature, and behaving like a State, is a much more possible thing. . . .

In its more developed phases I seem to see the New Republic as (if I may use an expressive bull) a sort of outspoken Secret Society, with which even the prominent men of the ostensible state may be openly affiliated. A vast number of men admit the need but hesitate at the means of revolution, and in this conception of a slowly growing new social order organised with open deliberation within the substance of the old, there are no doubt elements of technical treason, but an enormous gain in the thoroughness, efficiency, and stability of the possible change.

So it is, or at least in some such ways, that I conceive the growing sense of itself which the new class of modern efficients will develop, will become manifest in movements and concerns that are now heterogeneous and distinct, but will presently drift into co-operation and coalescence. This idea of a synthetic reconstruction within the bodies of the English-speaking States may very possibly clothe itself in quite other formulæ than my phrase of the New Republic; but the need is with us, the social elements are developing among us, the appliances are arranging themselves for the hands that will use them, and I cannot but believe that the idea of a spacious common action will presently come. In a few years I believe many men who are now rather aimless–men who have disconsolately watched the collapse of the old Liberalism–will be clearly telling themselves and one another of their adhesion to this new ideal. They will be working in schools and newspaper offices, in foundries and factories, in colleges and laboratories, in county councils and on school-boards–even, it may be, in pulpits–for the time when the coming of the New Republic will be ripe. It may be dawning even in the schools of law, because presently there will be a new and scientific handling of jurisprudence. The highly educated and efficient officers' mess will rise mechanically and drink to the Monarch, and sit down to go on discussing the New Republic's growth. I do not see, indeed, why an intelligent monarch himself, in these days, should not waive any silliness about Divine Right, and all the ill-bred pretensions that sit so heavily on a gentlemanly King, and come into the movement with these others. When the growing conception touches, as in America it has already touched, the legacy-leaving class, there will be fewer new Asylums perhaps, but more university chairs. . . .

So it is I conceive the elements of the New Republic taking shape and running together through the social mass, picking themselves out more and more clearly, from the shareholder, the parasitic speculator and the wretched multitudes of the Abyss. The New Republicans will constitute an informal and open freemasonry. In all sorts of ways they will be influencing and controlling the apparatus of the ostensible governments, they will be pruning irresponsible property, checking speculators and controlling the abyssward drift, but at that, at an indirect control, at any sort of fiction, the New Republic, from the very nature of its cardinal ideas, will not rest. The clearest and simplest statement, the clearest and simplest method, is inevitably associated with the conceptions of that science upon which the New Republic will arise. There will be a time, in peace it may be, or under the stresses of warfare, when the New Republic will find itself ready to arrive, when the theory will have been worked out and the details will be generally accepted, and the new order will be ripe to begin. And then, indeed, it will begin. What life or strength will be left in the old order to prevent this new order beginning?

IX. THE FAITH, MORALS, AND PUBLIC POLICY OF THE NEW REPUBLIC

IF the surmise of a developing New Republic–a Republic that must ultimately become a World State of capable, rational men, developing amidst the fading contours and colours of our existing nations and institutions–be indeed no idle dream, but an attainable possibility in the future, and to that end it is that the preceding Anticipations have been mainly written, it becomes a speculation of very great interest to forecast something of the general shape and something even of certain details of that common body of opinion which the New Republic, when at last it discovers and declares itself, will possess. Since we have supposed this New Republic will already be consciously and pretty freely controlling the general affairs of humanity before this century closes, its broad principles and opinions must necessarily shape and determine that still ampler future of which the coming hundred years is but the opening phase. There are many processes, many aspects of things, that are now, as it were, in the domain of natural laws and outside human control, or controlled unintelligently and superstitiously, that in the future, in the days of the coming New Republic, will be definitely taken in hand as part of the general work of humanity, as indeed already, since the beginning of the nineteenth century, the control of pestilences has been taken in hand. And in particular, there are certain broad questions much under discussion to which, thus far, I have purposely given a value disproportionately small:–

While the New Republic is gathering itself together and becoming aware of itself, that other great element, which I have called the People of the Abyss, will also have followed out its destiny. For many decades that development will be largely or entirely out of all human control. To the multiplying rejected of the white and yellow civilisations there will have been added a vast proportion of the black and brown races, and collectively those masses will propound the general question, "What will you do with us, we hundreds of millions, who cannot keep pace with you?" If the New Republic emerges at all it will emerge by grappling with this riddle; it must come into existence by the passes this Sphinx will guard. Moreover, the necessary results of the reaction of irresponsible wealth upon that infirm and dangerous thing the human will, the spreading moral rot of gambling which is associated with irresponsible wealth, will have been working out, and will continue to work out, so long as there is such a thing as irresponsible wealth pervading the social body. That too the New Republic must in its very development overcome. In the preceding chapter it is clearly implicit that I believe that the New Republic, as its consciousness and influence develop together, will meet, check, and control these things; but the broad principles upon which the control will go, the nature of the methods employed, still remain to be deduced. And to make that deduction, it is necessary that the primary conception of life, the fundamental, religious, and moral ideas of these predominant men of the new time should first be considered.

Now, quite inevitably, these men will be religious men. *Being themselves, as by the nature of the forces that have selected them they will certainly be, men of will and purpose, they will be disposed to find, and consequently they will find, an effect of purpose in the totality of things.* Either one must believe the Universe to be one and systematic, and held together by some omnipresent quality, or one must believe it to be a casual aggregation, an incoherent accumulation with no unity whatsoever outside the unity of the personality regarding it. All science and most modern religious systems presuppose the former, and to believe the former is, to anyone not too anxious to quibble, to believe in God. But I believe that these prevailing men of the future, like many of the saner men of to-day, having so formulated their fundamental belief, will presume to no knowledge whatever, will presume to no possibility of knowledge of the real being of God. They will have no positive definition of God at all. They will certainly not indulge in "that something, not ourselves, that makes for righteousness" (not defined) or any defective claptrap of that sort. They will content themselves with denying the self-contradictory absurdities of an obstinately anthropomorphic theology,* they will regard the whole of being, within themselves and without, as the sufficient revelation or God to their souls, and they will set themselves simply to that revelation, seeking its meaning towards themselves faithfully and courageously. Manifestly the essential being of man in this life is his will; he exists consciously only to *do;* his main interest in life is the choice between alternatives; and, since he moves through space and time to effects and consequences, a general purpose in space and time is the limit of his understanding. He can know God only under the semblance of a pervading purpose, of which his own individual freedom of will is a part, but he can understand that the purpose that exists in space and time is no more God than a voice calling out of impenetrable darkness is a man. To men of the kinetic type belief in God so manifest as purpose is irresistible, and, to all lucid minds, the being of God, save as that general atmosphere of imperfectly apprehended purpose in which our individual wills operate, is incomprehensible. To cling to any belief more detailed than this, to define and limit God in order to take hold of Him, to detach oneself and parts of the universe from God in some mysterious way in order to reduce life to a dramatic antagonism, is not faith, but infirmity. Excessive strenuous belief is not faith. By faith we disbelieve, and it is the drowning man, and not the strong swimmer, who clutches at the floating straw. It is in the nature of man, it is in the present purpose of things, that the real world of our experience and will should appear to us not only as a progressive existence

*As, for example, that God is an omniscient mind. This is the last vestige of that barbaric theology which regarded God as a vigorous but uncertain old gentleman with a bear and an inordinate lust for praise and propitiation. The modern idea is, indeed, scarcely more reasonable than the one it has replaced. A mind thinks, and feels, and wills; it passes from phase to phase; thinking and willing are a succession of mental states which follow and replace one another. But omniscience is a complete knowledge, not only of the present state, but of all past and future states, and, since it is all there at any moment, it cannot conceivably pass from phase to phase, it is stagnant, infinite, and eternal. An omniscient mind is as impossible, therefore, as an omnipresent moving body. God is outside our mental scope; only by faith can we attain Him; our most lucid moments serve only to render clearer His inaccessibility to our intelligence. We stand a little way up in a scale of existences that may, indeed, point towards Him, but can never bring Him to our scope. As the fulness of the conscious mental existence of a man stands to the subconscious activities of an amœba or of a visceral ganglion cell, so our reason forces us to admit other possible mental existences may stand to us. But such an existence, inconceivably great as it would be to us, would be scarcely nearer that transcendental God in whom the serious men of the future will, as a class, believe.

in space and time, but as a scheme of good and evil. But choice, the antagonism of good and evil, just as much as the formulation of things in space and time, is merely a limiting condition of human being, and in the thought of God as we conceive of Him in the light of faith, this antagonism vanishes. God is no moralist, God is no partisan; He comprehends and cannot be comprehended, and our business is only with so much of His purpose as centres on our individual wills.

So, or in some such phrases, I believe, these men of the New Republic will formulate their relationship to God. They will live to serve this purpose that presents Him, without presumption and without fear. For the same spacious faith that will render the idea of airing their egotisms in God's presence through prayer, or of any such quite personal intimacy, absurd, will render the idea of an irascible and punitive Deity ridiculous and incredible. . . .

The men of the New Republic will hold and understand quite clearly the doctrine that in the real world of man's experience, there is Free Will. They will understand that constantly, as a very condition of his existence, man is exercising choice between alternatives, and that a conflict between motives that have different moral values constantly arises. That conflict between Predestination and Free Will, which is so puzzling to untrained minds, will not exist for them. They will know that in the real world of sensory experience, will is free, just as new-sprung grass is green, wood hard, ice cold, and toothache painful. In the abstract world of reasoning science there is no green, no colour at all, but certain lengths of vibration; no hardness, but a certain reaction of molecules; no cold and no pain, but certain molecular consequences in the nerves that reach the misinterpreting mind. In the abstract world of reasoning science, moreover, there is a rigid and inevitable sequence of cause and effect; every act of man could be foretold to its uttermost detail, if only we knew him and all his circumstances fully; in the abstract world of reasoned science all things exist now potentially down to the last moment of infinite time. But the human will does not exist in the abstract world of reasoned science, in the world of atoms and vibrations, that rigidly predestinate scheme of things in space and time. The human will exists in the world of men and women, in this world where the grass is green and desire beckons, and the choice is often so wide and clear between the sense of what is desirable and what is more widely and remotely right. In this world of sense and the daily life, these men will believe with an absolute conviction, that there is free will and a personal moral responsibility in relation to that indistinctly seen purpose which is the sufficient revelation of God to them so far as this sphere of being goes. . . .

The conception they will have of that purpose will necessarily determine their ethical scheme. It follows manifestly that if we do really believe in Almighty God, the more strenuously and successfully we seek in ourselves and His world to understand the order and progress of things, and the more clearly we apprehend His purpose, the more assured and systematic will our ethical basis become.

If, like Huxley, we do not positively believe in God, then we may still cling to an ethical system which has become an organic part of our lives and habits, and finding it manifestly in conflict with the purpose of things, speak of the non-ethical order of the universe. But to anyone whose mind is pervaded by faith in God, a non-ethical universe in conflict with the incomprehensibly ethical soul of the Agnostic, is as incredible as a black-horned devil, an active, material anti-god with hoofs, tail, pitchfork, and Dunstan-scorched nose complete. To believe completely in God is to believe in the final rightness of all being. The ethical system that condemns the ways of life as wrong, or points to the ways of death as right, that countenances what the scheme of things condemns, and condemns the general purpose in things as it is now revealed to us, must prepare to follow the theological edifice upon which it was originally based. If the universe is non-ethical by our present standards, we must reconsider these standards and reconstruct our ethics. To hesitate to do so, however severe the conflict with old habits and traditions and sentiments may be, is to fall short of faith.

Now, so far as the intellectual life of the world goes, this present time is essentially the opening phase of a period of ethical reconstruction, a reconstruction of which the New Republic will possess the matured result. Throughout the nineteenth century there has been such a shattering and recasting of fundamental ideas, of the preliminaries to ethical propositions, as the world has never seen before. This breaking down and routing out of almost all the cardinal assumptions on which the minds of the Eighteenth Century dwelt securely, is a process akin to, but independent of, the development of mechanism, whose consequences we have traced. It is a part of that process of vigorous and fearless criticism which is the reality of science, and of which the development of mechanism and all that revolution in physical and social conditions we have been tracing, is merely the vast, imposing, material by-product. At present, indeed, its more obvious aspect on the moral and ethical side is destruction, anyone can see the chips flying, but it still demands a certain faith and patience to see the form that ensues. But it is not destruction, any more than a sculptor's work is stone-breaking.

The first chapter in the history of this intellectual development, its definite and formal opening, coincides with the opening of the nineteenth century and the publication of Malthus's *Essay on Population*. Malthus is one of those cardinal figures in intellectual history who state definitely for all time, things apparent enough after their formulation, but never effectively conceded before. He brought clearly and emphatically into the sphere of discussion a vitally important issue that had always been shirked and tabooed heretofore, the fundamental fact that the main mass of the business of human life centres about reproduction. He stated in clear, hard, decent, and unavoidable argument what presently Schopenhauer was to discover and proclaim, in language, at times, it would seem, quite unfitted for translation into English. And, having made his statement, Malthus left it, in contact with its immediate results.

Probably no more shattering book than the *Essay on Population* has ever been, or ever will be, written. It was aimed at the facile Liberalism of the Deists and Atheists of the eighteenth century; it made as clear as daylight that all forms of social reconstruction, all dreams of earthly golden ages must be either futile or insincere or both, until the problems of human increase were manfully faced. It proffered no suggestions for facing them (in spite of the unpleasant associations of Malthus's name), it aimed simply to wither the Rationalistic Utopias of the time and by anticipation, all the Communisms, Socialisms, and Earthly Paradise movements that have since been so abundantly audible in the world. That was its aim and its immediate effect. Incidentally it must have been a torturing soul-trap for innumerable idealistic but intelligent souls. Its indirect effects have been altogether greater. Aiming at unorthodox dreamers, it has set such forces in motion as have destroyed the very root-ideas of orthodox righteousness in the western world. Impinging on geological discovery, it awakened almost simultaneously in the minds of Darwin and Wallace, that train of thought that found expression and demonstration at last in the theory of natural selection. As that theory has been more and more thoroughly assimilated and understood by the general mind, it has destroyed, quietly but entirely, the belief in human equality which is implicit in all the "Liberalising" movements of the world. In the place of an essential equality, distorted only by tradition and early training, by the artifices of those devils of the Liberal cosmogony, "kingcraft" and "priestcraft," an equality as little affected by colour as the equality of a black chess pawn and a white, we discover that all men are individual and unique, and, through long ranges of comparison, superior and inferior upon countless scores. It has become apparent that whole masses of human population are, as a whole, inferior in their claim upon the future to other masses, that they cannot be given opportunities or trusted with power as the superior peoples are trusted, that their characteristic weaknesses are contagious and detrimental in the civilising fabric, and that their range of incapacity tempts and demoralises the strong. To give them equality is to sink to their level, to protect and cherish them is to be swamped in their fecundity. The confident and optimistic Radicalism of the earlier nineteenth century, and the humanitarian philanthropic type of Liberalism, have bogged themselves beyond hope in these realisations. The Socialist has shirked them as he has shirked the older crux of Malthus. Liberalism is a thing of the past, it is no longer a doctrine, but a faction. There must follow some new-born thing.

And as effectually has the mass of criticism that centres about Darwin destroyed the dogma of the Fall upon which the whole intellectual fabric of Christianity rests. For without a Fall there is no redemption, and the whole theory and meaning of the Pauline system is vain. In conjunction with the wide vistas opened by geological and astronomical discovery, the nineteenth century has indeed lost the very habit of thought from which the belief in a Fall arose. It is as if a hand had been put upon the head of the thoughtful man and had turned his eyes about from the past to the future. In matters of intelligence, at least, if not yet in matters of ethics and conduct, this turning round has occurred. In the past thought was legal in its spirit, it deduced the present from pre-existing prescription, it derived everything from the offences and promises of the dead; the idea of a universe of expiation was the most natural theory amidst such processes. The purpose the older theologians saw in the world was no more than the revenge–accentuated by the special treatment of a favoured minority–of a mysteriously incompetent Deity exasperated by an unsatisfactory creation. But modern thought is altogether too constructive and creative to tolerate such a conception, and in the vaster past that has opened to us, it can find neither offence nor promise, only a spacious scheme of events, opening out–perpetually opening out–with a quality of final purpose as irresistible to most men's minds as it is incomprehensible, opening out with all that inexplicable quality of design that, for example, some great piece of music, some symphony of Beethoven's, conveys. We see future beyond future and past behind past. It has been like the coming of dawn, at first a colourless dawn, clear and spacious, before which the mists whirl and fade, and there opens to our eyes not the narrow passage, the definite end we had imagined, but the rocky, ill-defined path we follow high amidst this limitless prospect of space and time. At first the dawn is cold–there is, at times, a quality of terror almost in the cold clearness of the morning twilight; but insensibly its coldness passes, the sky is touched with fire, and presently, up out of the dayspring in the east, the sunlight will be pouring. . . . And these men of the New Republic will be going about in the daylight of things assured.

And men's concern under this ampler view will no longer be to work out a system of penalties for the sins of dead men, but to understand and participate in this great development that now dawns on the human understanding. The insoluble problems of pain and death, gaunt, incomprehensible facts as they were, fall into place in the gigantic order that evolution unfolds. All things are integral in the mighty scheme, the slain builds up the slayer, the wolf grooms the horse into swiftness, and the tiger calls for wisdom and courage out of man. All things are integral, but it has been left for men to be consciously integral, to take, at last, a share in the process, to have wills that have caught a harmony with the universal will, as sand-grains flash into splendour under the blaze of the sun. There will be many who will never be called to this religious conviction, who will lead their little lives like fools, playing foolishly with religion and all the great issues of life, or like the beasts that perish, having sense alone; but those who, by character and intelligence, are predestinate to participate in the reality of life, will fearlessly shape all their ethical determinations and public policy anew, from a fearless study of themselves and the apparent purpose that opens out before them.

Very much of the cry for faith that sounds in contemporary life so loudly, and often with so distressing a note of sincerity, comes from the unsatisfied egotisms of unemployed, and, therefore, unhappy and craving people; but much is also due to the distress in the minds of active and serious men, due to the conflict of inductive knowledge, with conceptions of right and wrong deduced from unsound, but uncriticised, first principles. The old ethical principles, the principle of equivalents or justice, the principle of self-sacrifice, the various vague and arbitrary ideas of purity, chastity, and sexual "sin," came like rays out of the theological and philosophical lanterns men carried in the darkness. The ray of the lantern indicated and directed and one followed it as one follows a path. But now there has come a new view of man's place in the scheme of time and space, a new illumination, dawn; the lantern-rays fade in the growing brightness, and the lanterns that shone so brightly are becoming smoky and dim. To many men this is no more than a waning of the lanterns, and they call for new ones, or a trimming of the old. They blame the day for putting out these flares. And some go apart, out of the glare of life, into corners of obscurity, where the radiation of the lantern may still be faintly traced. But, indeed, with the new light there has come the time for new methods; the time of lanterns, the time of deductions from arbitrary first principles is over. The act of faith is no longer to follow your lantern, but to put it down. We can see about us, and by the landscape we must go.*

*It is an interesting by-way from our main thesis to speculate on the spiritual pathology of the functionless wealthy, the half-educated independent women of the middle class, and the people of the Abyss. While the segregating new middle class, whose religious and moral development forms our main interest, is developing its spacious and confident Theism, there will, I imagine, be a steady decay in the various Protestant congregations. They have played a noble part in the history of the world, their spirit will live for ever, but their formulæ and organisation wax old like a garment. Their moral austerity–that touch of contempt for the unsubstantial æsthetic, which has always distinguished Protestantism–is naturally repellent to the irresponsible rich and to artistic people of the weaker type, and the face of Protestantism has ever been firm even to hardness against the self-indulgent, the idler, and the prolific, useless poor. The rich as a class and the people of the Abyss, so far as they move towards any existing religious body, will be attracted by the moral kindliness, the picturesque organisation and venerable tradition of the Roman Catholic Church. We are only in the very beginning of a great Roman Catholic revival. The diversified country-side of the coming time will show many a splendid cathedral, many an elaborate monastic palace, towering amidst the abounding colleges and technical schools. Along the moving platforms of the urban centre, and athwart the shining advertisements that will adorn them, will go the ceremonial procession, all glorious with banners and censer-bearers, and the meek, blue-shaven priests and barefooted, rope-girdled, holy men. And the artful politician of the coming days, until the broom of the New Republic sweep him up, will arrange the miraculous planks of his platform always with an eye upon the priest. Within the ample sheltering arms of the Mother Church many eccentric cults will develop. The curious may study the works of M. Huysmans to learn of the mystical propitiation of God, Who made heaven and earth, by the bed-sores of hysterical girls. The future as I see it swarms with Durtals and Sister Teresas; countless ecstatic nuns, holding their Maker as it were *in deliciæ,* will shelter from the world in simple but costly refuges of refined austerity. Where miracles are needed, miracles will occur.

Except for a few queer people, nourished on "Maria Monk" and suchlike anti-papal pornography, I doubt if there will be any Protestants left among the

How will the landscape shape itself to the dominant men of the new time and in relation to themselves? What is the will and purpose that these men of will and purpose will find above and comprehending their own? Into this our inquiry resolves itself. They will hold with Schopenhauer, I believe, and with those who build themselves on Malthus and Darwin, that the scheme of being in which we live is a struggle of existences to expand and develop themselves to their full completeness, and to propagate and increase themselves. But, being men of action, they will feel nothing of the glamour of misery that irresponsible and sexually vitiated shareholder, Schopenhauer, threw over this recognition. The final object of this struggle among existences they will not understand; they will have abandoned the search for ultimates; they will state this scheme of a struggle as a proximate object, sufficiently remote and spacious to enclose and explain all their possible activities. They will seek God's purpose in the sphere of their activities, and desire no more, as the soldier in

irresponsible rich. Those who do not follow the main current will probably take up with weird science-denouncing sects of the faith-healing type, or with such pseudo-scientific gibberish as Theosophy. Mrs. Piper (in an inelegant attitude and with only the whites of her eyes showing) has restored the waning faith of Professor James in human immortality, and I do not see why that lady should stick at one dogma amidst the present quite insatiable demand for creeds. Shintoism and either a cleaned or, more probably, a scented Obi, might in vigorous hands be pushed to a very considerable success in the coming years; and I do not see any absolute impossibility in the idea of an after-dinner witch-smelling in Park Lane with a witch-doctor dressed in feathers. It might be made amazingly picturesque. People would attend it with an air of intellectual liberality, not, of course, believing in it absolutely, but admitting "there must be Something in it!" That Something in it! "The fool hath said in his heart, there is no God," and after that he is ready to do anything with his mind and soul. It is by faith we disbelieve.

And, of course, there will be much outspoken Atheism and Anti-religion of the type of the Parisian Devil-Worship imbecilities. Young men of means will determine to be "wicked." They will do silly things that will strike them as being indecent and blasphemous and dreadful–black masses and suchlike nonsense–and then they will get scared. The sort of thing it will be to shock orthodox maiden aunts and make Olympus ring with laughter. A taking sort of nonsense already loose, I find, among very young men is to say, "Understand, I am non-moral." Two thoroughly respectable young gentlemen coming from quite different circles have recently introduced their souls to me in this same formula. Both, I rejoice to remark, are married, both are steady and industrious young men, trustworthy in word and contract, dressed in accordance with current conceptions, and behaving with perfect decorum. One, no doubt for sinister ends, aspires to better the world through a Socialistic propaganda. That is all. But in a tight corner some day that silly little formula may just suffice to trip up one or other of these men. To many of the irresponsible rich, however, that little "Understand, I am non-moral" may prove of priceless worth.

battle desires no more, than the immediate conflict before him. They will admit failure as an individual aspect of things, as a soldier seeking victory admits the possibility of death; but they will refuse to admit as a part of their faith in God that any existence, even if it is an existence that is presently entirely erased, can be needless or vain. It will have reacted on the existences that survive; it will be justified for ever in the modification it has produced in them. They will find in themselves–it must be remembered I am speaking of a class that has naturally segregated, and not of men as a whole–a desire, a passion almost, to create and organise, to put in order, to get the maximum result from certain possibilities. They will all be artists in reality, with a passion for simplicity and directness and an impatience of confusion and inefficiency. The determining frame of their ethics, the more spacious scheme to which they will shape the schemes of their individual wills, will be the elaboration of that future world-state to which all things are pointing. They will not conceive of it as a millennial paradise, a blissful inconsequent stagnation, but as a world-state of active ampler human beings, full of knowledge and energy, free from much of the baseness and limitations, the needless pains and dishonours of the world disorder of to-day, but still struggling, struggling against ampler but still too narrow restrictions and for still more spacious objects than our vistas have revealed. For that as a general end, for the special work that contributes to it as an individual end, they will make the plans and the limiting rules of their lives.

It is manifest that a reconstructed ethical system, reconstructed in the light of modern science and *to meet the needs of such temperaments and characters as the evolution of mechanism will draw together and develop* will give very different values from those given by the existing systems (if they can be called systems) to almost all the great matters of conduct. Under scientific analysis the essential facts of life are very clearly shown to be two–birth and death. All life is the effort of the thing born, driven by fears, guided by instincts and desires, to evade death, to evade even the partial death of crippling or cramping or restriction, and to attain to effective procreation, to the victory of another birth. Procreation is the triumph of the living being over death; and in the case of man, who adds mind to this body, it is not only in his child but in the dissemination of his thought, the expression of his mind in things done and made, that his triumph is to be found. And the ethical system of these men of the New Republic, the ethical system which will dominate the world-state, will be shaped primarily to favour the procreation of what is fine and efficient and beautiful in humanity–beautiful and strong bodies, clear and powerful minds, and a growing body of knowledge–and to check the procreation of base and servile types, of fear-driven and cowardly souls, of all that is mean and ugly and bestial in the souls, bodies, or habits of men. To do the latter is to do the former; the two things are inseparable. And the method that nature has followed hitherto in the shaping of the world, whereby weakness was prevented from propagating weakness, and cowardice and feebleness were saved from the accomplishment of their desires, the method that has only one alternative, the method that must in some cases still be called in to the help of man, is death. In the new vision death is no inexplicable horror, no pointless terminal terror to the miseries of life, it is the end of all the pain of life, the end of the bitterness of failure, the merciful obliteration of weak and silly and pointless things. . . .

The new ethics will hold life to be a privilege and a responsibility, not a sort of night refuge for base spirits out of the void; and the alternative in right conduct between living fully, beautifully, and efficiently will be to die. For a multitude of contemptible and silly creatures, fear-driven and helpless and useless, unhappy or hatefully happy in the midst of squalid dishonour, feeble, ugly, inefficient, born of unrestrained lusts, and increasing and multiplying through sheer incontinence and stupidity, the men of the New Republic will have little pity and less benevolence. To make life convenient for the breeding of such people will seem to them not the most virtuous and amiable thing in the world, as it is held to be now, but an exceedingly abominable proceeding. Procreation is an avoidable thing for sane persons of even the most furious passions, and the men of the New Republic will hold that the procreation of children who, by the circumstances of their parentage, *must* be diseased bodily or mentally–I do not think it will be difficult for the medical science of the coming time to define such circumstances–is absolutely the most loathsome of all conceivable sins. They will hold, I anticipate, that a certain portion of the population–the small minority, for example, afflicted with indisputably transmissible diseases, with transmissible mental disorders, with such hideous incurable habits of mind as the craving for intoxication–exists only on sufferance, out of pity and patience, and on the understanding that they do not propagate; and I do not foresee any reason to suppose that they will hesitate to kill when that sufferance is abused. And I imagine also the plea and proof that a grave criminal is also insane will be regarded by them not as a reason for mercy, but as an added reason for death. I do not see how they can think otherwise on the principles they will profess.

The men of the New Republic will not be squeamish, either, in facing or inflicting death, because they will have a fuller sense of the possibilities of life than we possess. They will have an ideal that will make killing worth the while; like Abraham, they will have the faith to kill, and they will have no superstitions about death. They will naturally regard the modest suicide of incurably melancholy, or diseased or helpless persons as a high and courageous act of duty rather than a crime. And since they will regard, as indeed all men raised above a brutish level do regard, a very long term of imprisonment as infinitely worse than death, as being, indeed, death with a living misery added to its natural terror, they will, I conceive, where the whole tenor of a man's actions, and not simply some incidental or impulsive action, seems to prove him unfitted for free life in the world, consider him carefully, and condemn him, and remove him from being. All such killing will be done with an opiate, for death is too grave a thing to be made painful or dreadful, and used as a deterrent from crime. If deterrent punishments are used at all in the code of the future, the deterrent will neither be death, nor mutilation of the body, nor mutilation of the life by imprisonment, nor any horrible things like that, but good, scientifically caused pain, that will leave nothing but a memory. Yet even the memory of overwhelming pain is a sort of mutilation of the soul. The idea that only those who are fit to live freely in an orderly world-state should be permitted to live, is entirely against the use of deterrent punishments at all. Against outrageous conduct to children or women, perhaps, or for very cowardly or brutal assaults of any sort, the men of the future may consider pain a salutary remedy, at least during the ages of transition while the brute is still at large. But since most acts of this sort done under conditions that neither torture nor exasperate, point to an essential vileness in the perpetrator, I am inclined to think that even in these cases the men of the coming time will be far less disposed to torture than to kill. They will have another aspect to consider. The conscious infliction of pain *for the sake of the pain* is against the better nature of man, and it is unsafe and demoralising for anyone to undertake this duty. To kill under the seemly conditions science will afford is a far less offensive thing. The rulers of the future will grudge making good people into jailers, warders, punishment-dealers, nurses, and attendants on the bad. People who cannot live happily and freely in the world without spoiling the lives of others are better out of it. That is a current sentiment even to-day, but the men of the New Republic will have the courage of their opinions.

And the type of men that I conceive emerging in the coming years will deal simply and logically not only with the business of death, but with birth. At present the sexual morality of the civilised world is the most illogical and incoherent system of wild permissions and insane prohibitions, foolish tolerance and ruthless cruelty that it is possible to imagine. Our current civilisation is a sexual lunatic. And it has lost its reason in this respect under the stresses of the new birth of things, largely through the difficulties that have stood in the way, and do still, in a diminishing degree, stand in the way of any sane discussion of the matter as a whole. To approach it is to approach excitement. So few people seem to be leading happy and healthy sexual lives that to mention the very word "sexual" is to set them stirring, to brighten the eye, lower the voice, and blanch or flush the cheek with a flavour of guilt. We are all, as it were, keeping our secrets and hiding our shames. One of the most curious revelations of this fact occurred only a few years ago, when the artless outpourings in fiction of certain young women who had failed to find light on problems that pressed upon them for solution (and which it was certainly their business as possible wives and mothers to solve) roused all sorts of respectable people to a quite insane vehemence of condemnation. Now, there are excellent reasons and a permanent necessity for the preservation of decency, and for a far more stringent suppression of matter that is merely intended to excite than at present obtains, and the chief of these reasons lies in the need of preserving the young from a premature awakening, and indeed, in the interests of civilisation, in positively delaying the period of awakening, retarding maturity and lengthening the period of growth and preparation as much as possible. But purity and innocence may be prolonged too late; innocence is really no more becoming to adults than a rattle or a rubber consoler, and the bashfulness that hampers this discussion, that permits it only in a furtive, silly sort of way, has its ugly consequences in shames and cruelties, in miserable households and pitiful crises, in the production of countless, needless, and unhappy lives. Indeed, too often we carry our decency so far as to make it suggestive and stimulating in a non-natural way; we invest the plain business of reproduction with a mystic religious quality far more unwholesome than a savage nakedness could possibly be.

The essential aspect of all this wild and windy business of the sexual relations is, after all, births. Upon this plain fact the people of the emergent New Republic will unhesitatingly go. The pre-eminent value of sexual questions in morality lies in the fact that the lives which will constitute the future are involved. If they are not involved, if we can dissociate this relationship from this issue, then sexual questions become of no more importance than the morality of one's deportment at chess, or the general morality of outdoor games. Indeed, then the question of sexual relationships would be entirely on all fours with, and probably very analogous to, the question of golf. In each case it would be for the medical man and the psychologist to decide how far the thing was wholesome and permissible, and how far it was an aggressive bad habit and an absorbing waste of time and energy. An able-bodied man continually addicted to love-making that had no result in offspring would be just as silly and morally objectionable as an able-bodied man who devoted his chief energies to hitting little balls over golf-links. But no more. Both would probably be wasting the lives of other human beings–the golfer must employ his caddie. It is entirely the matter of births, and a further consideration to be presently discussed, that makes this analogy untrue. It does not, however, make it so untrue as to do away with the probability that in many cases the emergent men of the new time will consider sterile gratification a moral and legitimate thing. St. Paul tells us that it is better to marry than to burn, but to beget children on that account will appear, I imagine, to these coming men as an absolutely loathsome proceeding. They will stifle no spread of knowledge that will diminish the swarming misery of childhood in the slums, they will regard the disinclination of the witless "Society" woman to become a mother as a most amiable trait in her folly. In our bashfulness about these things we talk an abominable lot of nonsense; all this uproar one hears about the Rapid Multiplication of the Unfit and the future of the lower races takes on an entirely different complexion directly we face known, if indelicate, facts. Most of the human types, that by civilised standards are undesirable, are quite willing to die out through such suppressions if the world will only encourage them a little. They multiply in sheer ignorance, but they do not desire multiplication even now, and they can easily be made to dread it. Sensuality aims not at life, but at itself. I believe that the men of the New Republic will deliberately shape their public policy along these lines. They will rout out and illuminate urban rookeries and all places where the base can drift to multiply; they will contrive a land legislation that will keep the black, or yellow, or mean-white squatter on the move; they will see to it that no parent can make a profit out of a child, so that childbearing shall cease to be a hopeful speculation for the unemployed poor; and they will make the maintenance of a child the first charge upon the parents who have brought it into the world. Only in this way can progress escape being clogged by the products of the security it creates. The development of science has lifted famine and pestilence from the shoulders of man, and it will yet lift war–for some other end than to give him a spell of promiscuous and finally cruel and horrible reproduction.

No doubt the sentimentalist and all whose moral sense has been vigorously trained in the old school will find this rather a dreadful suggestion; it amounts to saying that for the Abyss to become a "hotbed" of sterile immorality will fall in with the deliberate policy of the ruling class in the days to come. At anyrate, it will be a terminating evil. At present the Abyss is a hotbed breeding undesirable and too often fearfully miserable children. *That* is something more than a sentimental horror. Under the really very horrible morality of to-day, the spectacle of a mean-spirited, under-sized, diseased little man, quite incapable of earning a decent living even for himself, married to some underfed, ignorant, ill-shaped, plain and diseased little woman, and guilty of the lives of ten or twelve ugly, ailing children, is regarded as an extremely edifying spectacle, and the two parents consider their reproductive excesses as giving them a distinct claim upon less fecund and more prosperous people. Benevolent persons throw themselves with peculiar ardour into a case of this sort, and quite passionate efforts are made to strengthen the mother against further eventualities and protect the children until they attain to nubile years. Until the attention of the benevolent persons is presently distracted by a new case. . . . Yet so powerful is the suggestion of current opinions that few people seem to see nowadays just what a horrible and criminal thing this sort of family, seen from the point of view of social physiology, appears.

And directly such principles as these come into effective operation, and I believe that the next hundred years will see this new phase of the human history beginning, there will recommence a process of physical and mental improvement in mankind, a raising and elaboration of the average man, that has virtually been in suspense during the greater portion of the historical period. It is possible that in the last hundred years, in the more civilised states of the world, the average of humanity has positively fallen. All philanthropists, all our religious teachers, seem to be in a sort of informal conspiracy to preserve an atmosphere of mystical ignorance about these matters, which, in view of the irresistible nature of the sexual impulse, results in a swelling tide of miserable little lives. Consider what it will mean to have perhaps half the population of the world, in every generation, restrained from or tempted to evade reproduction! This thing, this euthanasia of the weak and sensual, is possible. On the principles that will probably animate the predominant classes of the new time, it will be permissible, and I have little or no doubt that in the future it will be planned and achieved.

If birth were all the making of a civilised man, the men of the future, on the general principles we have imputed to them, would under no circumstances find the birth of a child, healthy in body and brain, more than the most venial of offences. But birth gives only the beginning, the raw material, of a civilised man. The perfect civilised man is not only a sound, strong body but a very elaborate fabric of mind. He is a fabric of moral suggestions that become mental habits, a magazine of more or less systematised ideas, a scheme of knowledge and training and an æsthetic culture. He is the child not only of parents but of a home and of an education. He has to be carefully guarded from physical and moral contagions. A reasonable probability of ensuring home and education and protection without any parasitic dependence on people outside the kin of the child, will be a necessary condition to moral birth under such general principles as we have supposed. Now, this sweeps out of reason any such promiscuity of healthy people as the late Mr. Grant Allen is supposed to have advocated–but, so far as I can understand him, did not. But whether it works out to the taking over of the permanent monogamic marriage of the old morality, as a going concern, is another matter. Upon this matter I must confess my views of the trend of things in the future do not seem to be finally shaped. The question involves very obscure physiological and psychological considerations. A man who aims to become a novelist naturally pries into these matters whenever he can, but the vital facts are very often hard to come by. It is probable that a great number of people could be paired off in couples who would make permanently happy and successful monogamic homes for their sound and healthy children. At anyrate, if a certain freedom of regrouping were possible within a time limit, this might be so. But I am convinced that a large proportion of married couples in the world to-day are not completely and happily matched, that there is much mutual limitation, mutual annulment and mutual exasperation. Home with an atmosphere of contention is worse than none for the child, and it is the interest of the child, and that alone, that will be the test of all these things. I do not think that the arrangement in couples is universally applicable, or that celibacy (tempered by sterile vice) should be its only alternative. Nor can I see why the union of two childless people should have an indissoluble permanence or prohibit an ampler grouping. The question is greatly complicated by the economic disadvantage of women, which makes wifehood the chief feminine profession, while only for an incidental sort of man is marriage a source of income, and further by the fact that most women have a period of maximum attractiveness after which it would be grossly unfair to cast them aside. From the point of view we are discussing, the efficient mother who can make the best of her children, is the most important sort of person in the state. She is a primary necessity to the coming civilisation. Can the wife in any sort of polygamic arrangement, or a woman of no assured status, attain to the maternal possibilities of the ideal monogamic wife? One is disposed to answer, No. But then, on the other hand, does the ordinary monogamic wife do that? We are dealing with the finer people of the future, strongly individualised people, who will be much freer from stereotyped moral suggestions and much less inclined to be dealt with wholesale than the people of to-day.

I have already shown cause in these Anticipations to expect a period of disorder and hypocrisy in matters of sexual morality. I am inclined to think that, when the New Republic emerges on the other side of this disorder, there will be a great number of marriage contracts possible between men and women, and that the strong arm of the State will insist only upon one thing–the security and welfare of the child. The inevitable removal of births from the sphere of an uncontrollable Providence to the category of deliberate acts, will enormously enhance the responsibility of the parent–and of the State that has failed to adequately discourage the philo-progenitiveness of the parent–towards the child. Having permitted the child to come into existence, public policy and the older standard of justice alike demand, under these new conditions, that it must be fed, cherished, and educated, not merely up to a respectable minimum, but to the full height of its possibilities. The State will, therefore, be the reserve guardian of all children. If they are being under-nourished, if their education is being neglected, the State will step in, take over the responsibility of their management, and enforce their charge upon the parents. The first liability of a parent will be to his child, and for his child; even the dues of that darling of our current law, the landlord, will stand second to that. This conception of the responsibility of the parents and the State to the child and the future runs quite counter to the general ideas of to-day. These general ideas distort grim realities. Under the most pious amiable professions, all the Christian states of to-day are, as a matter of fact, engaged in slave-breeding. The chief result, though of course it is not the intention, of the activities of priest and moralist to-day in these matters, is to lure a vast multitude of little souls into this world, for whom there is neither sufficient food, nor love, nor schools, nor any prospect at all in life but the insufficient bread of servitude. It is a result that endears religion and purity to the sweating employer, and leads unimaginative bishops, who have never missed a meal in their lives, and who know nothing of the indescribable bitterness of a handicapped entry into this world, to draw a complacent contrast with irreligious France. It is a result that must necessarily be recognised in its reality, and faced by these men who will presently emerge to rule the world; men who will have neither the plea of ignorance, nor moral stupidity, nor dogmatic revelation to excuse such elaborate cruelty.

And having set themselves in these ways to raise the quality of human birth, the New Republicans will see to it that the children who do at last effectually get born come into a world of spacious opportunity. The half-educated, unskilled pretenders, professing impossible creeds and propounding ridiculous curricula, to whom the unhappy parents of to-day must needs entrust the intelligences of their children; these heavy-handed barber-surgeons of the mind, these schoolmasters, with their ragtag and bobtail of sweated and unqualified assistants, will be succeeded by capable, self-respecting men and women, constituting the most important profession of the world. The windy pretences of "forming character," supplying moral training, and so forth, under which the educationalist of to-day conceals the fact that he is incapable of his proper task of training, developing and equipping the mind, will no longer be made by the teacher. Nor will the teacher be permitted to subordinate his duties to the entirely irrelevant business of his pupils' sports. The teacher will teach, and confine his moral training, beyond enforcing truth and discipline, to the exhibition of a capable person doing his duty as well as it can be done. He will know that his utmost province is only a part of the educational process, that equally important educational influences are the home and the world of thought about the pupil and himself. The whole world will be thinking and learning; the old idea of "completing" one's education will have vanished with the fancy of a static universe; every school will be a preparatory school, every college. The school and college will probably give only the keys and apparatus of thought, a necessary language or so, thoroughly done, a sound mathematical training, drawing, a wide and reasoned view of philosophy, some good exercises in dialectics, a training in the use of those stores of fact that science has made. So equipped, the young man and young woman will go on to the technical school of their chosen profession, and to the criticism of contemporary practice for their special efficiency, and to the literature of contemporary thought for their general development. . . .

And while the emergent New Republic is deciding to provide for the swarming inferiority of the Abyss, and developing the morality and educational system of the future, in this fashion, it will be attacking that mass of irresponsible property that is so unavoidable and so threatening under present conditions. The attack will, of course, be made along lines that the developing science of economics will trace in the days immediately before us. A scheme of death-duties and of heavy graduated taxes upon irresponsible incomes, with, perhaps, in addition, a system of terminable liability for borrowers, will probably suffice to control the growth of this creditor elephantiasis. The detailed contrivances are for the specialist to make. If there is such a thing as bitterness in the public acts of the New Republicans, it will probably be found in the measures that will be directed against those who are parasitic, or who attempt to be parasitic, upon the social body, either by means of gambling, by manipulating the medium of exchange, or by such interventions upon legitimate transactions as, for example, the legal trade-union in Great Britain contrives in the case of house property and land. Simply because he fails more often than he succeeds, there is still a disposition among sentimental people to regard the gambler or the speculator as rather a dashing, adventurous sort of person, and to contrast his picturesque gallantry with the sober certainties of honest men. The men of the New Republic will be obtuse to the glamour of such romance; they will regard the gambler simply as a mean creature who hangs about the social body in the hope of getting something for nothing, who runs risks to filch the possessions of other men, exactly as a thief does. They will put the two on a footing, and the generous gambler, like the kindly drunkard, in the face of their effectual provision for his little weakness, will cease to complain that his worst enemy is himself. And, in dealing with speculation, the New Republic will have the power of an assured faith and purpose, and the resources of an economic science that is as yet only in its infancy. In such matters the New Republic will entertain no superstition of *laissez faire*. Money and credit are as much human contrivances as bicycles, and as liable to expansion and modification as any other sort of prevalent but imperfect machine.

And how will the New Republic treat the inferior races? How will it deal with the black? how will it deal with the yellow man? how will it tackle that alleged termite in the woodwork, the Jew? Certainly not as races at all. It will aim to establish, and it will at last, though probably only after a second century has passed, establish a world-state with a common language and a common rule. All over the world its roads, its standards, its laws, and its apparatus of control will run. It will, I have said, make the multiplication of those who fall behind a certain standard of social efficiency unpleasant and difficult, and it will have cast aside any coddling laws to save adult men from themselves.* It will tolerate no dark corners where the people of the Abyss may fester, no vast diffused slums of peasant proprietors, no stagnant plague-preserves. Whatever men may come into its efficient citizenship it will let come–white, black, red, or brown; the efficiency will be the test. And the Jew also it will treat as any other man. It is said that the Jew is incurably a parasite on the apparatus of credit. If there are parasites on the apparatus of credit, that is a reason for the legislative cleaning of the apparatus of credit, but it is no reason for the special treatment of the Jew. If the Jew has a certain incurable tendency to social parasitism, and we make social parasitism impossible, we shall abolish the Jew, and if he has not, there is no need to abolish the Jew. We are much more likely to find we have abolished the Caucasian solicitor. I really do not understand the exceptional attitude people take up against the Jews. There is something very ugly about many Jewish faces, but there are Gentile faces just as coarse and gross. The Jew asserts himself in relation to his nationality with a singular tactlessness,

**Vide* Mr. Archdall Read's excellent and suggestive book, "The Present Evolution of Man."

but it is hardly for the English to blame that. Many Jews are intensely vulgar in dress and bearing, materialistic in thought, and cunning and base in method, but no more so than many Gentiles. The Jew is mentally and physically precocious, and he ages and dies sooner than the average European, but in that and in a certain disingenuousness he is simply on all fours with the short, dark Welsh. He foregathers with those of his own nation, and favours them against the stranger, but so do the Scotch. I see nothing in his curious, dispersed nationality to dread or dislike. He is a remnant and legacy of mediævalism, a sentimentalist, perhaps, but no furtive plotter against the present progress of things. He was the mediæval Liberal; his persistent existence gave the lie to Catholic pretensions all through the days of their ascendency, and to-day he gives the lie to all our yapping "nationalisms," and sketches in his dispersed sympathies the coming of the world-state. He has never been known to burke a school. Much of the Jew's usury is no more than social scavenging. The Jew will probably lose much of his particularism, intermarry with Gentiles, and cease to be a physically distinct element in human affairs in a century or so. But much of his moral tradition will, I hope, never die. . . . And for the rest, those swarms of black, and brown, and dirty-white, and yellow people, who do not come into the new needs of efficiency?

Well, the world is a world, not a charitable institution, and I take it they will have to go. The whole tenor and meaning of the world, as I see it, is that they have to go. So far as they fail to develop sane, vigorous, and distinctive personalities for the great world of the future, it is their portion to die out and disappear.

The world has a purpose greater than happiness; our lives are to serve God's purpose, and that purpose aims not at man as an end, but works through him to greater issues. . . . This, I believe, will be the distinctive quality of the New Republican's belief. And, for that reason, I have not even speculated whether he will hold any belief in human immortality or no. He will certainly not believe there is any *post mortem* state of rewards and punishments because of his faith in the sanity of God, and I do not see how he will trace any reaction between this world and whatever world there may be of disembodied lives. Active and capable men of all forms of religious profession to-day tend in practice to disregard the question of immortality altogether. So, to a greater degree, will the kinetic men of the coming time. We may find that issue interesting enough when we turn over the leaf, but at present we have not turned over the leaf. On this side, in this life, the relevancy of things points not in the slightest towards the immortality of our egotisms, but convergently and overpoweringly to the future of our race, to that spacious future, of which these weak, ambitious Anticipations are, as it were, the dim reflection seen in a shallow and troubled pool.

For that future these men will live and die.

THE END.

A CATALOG OF SELECTED

DOVER BOOKS

IN ALL FIELDS OF INTEREST

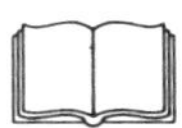

A CATALOG OF SELECTED DOVER BOOKS IN ALL FIELDS OF INTEREST

ANATOMY: A Complete Guide for Artists, Joseph Sheppard. A master of figure drawing shows artists how to render human anatomy convincingly. Over 460 illustrations. 224pp. 8⅜ x 11¼. 27279-6 Pa. $11.95

MEDIEVAL CALLIGRAPHY: Its History and Technique, Marc Drogin. Spirited history, comprehensive instruction manual covers 13 styles (ca. 4th century thru 15th). Excellent photographs; directions for duplicating medieval techniques with modern tools. 224pp. 8⅜ x 11¼. 26142-5 Pa. $12.95

DRIED FLOWERS: How to Prepare Them, Sarah Whitlock and Martha Rankin. Complete instructions on how to use silica gel, meal and borax, perlite aggregate, sand and borax, glycerine and water to create attractive permanent flower arrangements. 12 illustrations. 32pp. 5⅜ x 8½. 21802-3 Pa. $1.00

EASY-TO-MAKE BIRD FEEDERS FOR WOODWORKERS, Scott D. Campbell. Detailed, simple-to-use guide for designing, constructing, caring for and using feeders. Text, illustrations for 12 classic and contemporary designs. 96pp. 5⅜ x 8½. 25847-5 Pa. $2.95

SCOTTISH WONDER TALES FROM MYTH AND LEGEND, Donald A. Mackenzie. 16 lively tales tell of giants rumbling down mountainsides, of a magic wand that turns stone pillars into warriors, of gods and goddesses, evil hags, powerful forces and more. 240pp. 5⅜ x 8½. 29677-6 Pa. $6.95

THE HISTORY OF UNDERCLOTHES, C. Willett Cunnington and Phyllis Cunnington. Fascinating, well-documented survey covering six centuries of English undergarments, enhanced with over 100 illustrations: 12th-century laced-up bodice, footed long drawers (1795), 19th-century bustles, 19th-century corsets for men, Victorian "bust improvers," much more. 272pp. 5⅝ x 8¼. 27124-2 Pa. $9.95

ARTS AND CRAFTS FURNITURE: The Complete Brooks Catalog of 1912, Brooks Manufacturing Co. Photos and detailed descriptions of more than 150 now very collectible furniture designs from the Arts and Crafts movement depict davenports, settees, buffets, desks, tables, chairs, bedsteads, dressers and more, all built of solid, quarter-sawed oak. Invaluable for students and enthusiasts of antiques, Americana and the decorative arts. 80pp. 6½ x 9¼. 27471-3 Pa. $8.95

HOW WE INVENTED THE AIRPLANE: An Illustrated History, Orville Wright. Fascinating firsthand account covers early experiments, construction of planes and motors, first flights, much more. Introduction and commentary by Fred C. Kelly. 76 photographs. 96pp. 8¼ x 11. 25662-6 Pa. $8.95

THE ARTS OF THE SAILOR: Knotting, Splicing and Ropework, Hervey Garrett Smith. Indispensable shipboard reference covers tools, basic knots and useful hitches; handsewing and canvas work, more. Over 100 illustrations. Delightful reading for sea lovers. 256pp. 5⅜ x 8½. 26440-8 Pa. $7.95

FRANK LLOYD WRIGHT'S FALLINGWATER: The House and Its History, Second, Revised Edition, Donald Hoffmann. A total revision–both in text and illustrations–of the standard document on Fallingwater, the boldest, most personal architectural statement of Wright's mature years, updated with valuable new material from the recently opened Frank Lloyd Wright Archives. "Fascinating"–*The New York Times*. 116 illustrations. 128pp. 9¼ x 10¾. 27430-6 Pa. $11.95